T0182144

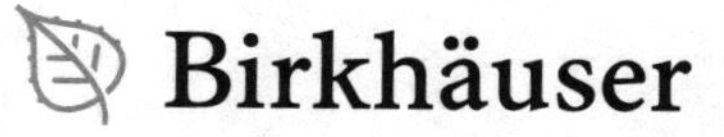
Birkhäuser

Systems & Control: Foundations & Applications

More information about this series at http://www.springer.com/series/4895

Naci Saldi • Tamás Linder • Serdar Yüksel

Finite Approximations in Discrete-Time Stochastic Control

Quantized Models and Asymptotic Optimality

Naci Saldi
Department of Natural and Mathematical Sciences
Ozyegin University
Istanbul, Turkey

Tamás Linder
Department of Mathematics and Statistics
Queen's University
Kingston, Ontario, Canada

Serdar Yüksel
Department of Mathematics & Statistics
Queen's University
Kingston, Ontario, Canada

ISSN 2324-9749 ISSN 2324-9757 (electronic)
Systems & Control: Foundations & Applications
ISBN 978-3-030-07710-5 ISBN 978-3-319-79033-6 (eBook)
https://doi.org/10.1007/978-3-319-79033-6

Mathematics Subject Classification (2010): 93E20, 90C40, 60J20, 49J55, 90B99, 60J05

Printed on acid-free paper

This book is published under the imprint Birkhäuser, www.birkhauser-science.com by the registered company Springer International Publishing AG part of Springer Nature.
The registered company address is: Gewerbestrasse 11, 6330 Cham, Switzerland

Contents

Chapter 1
Introduction and Summary

1.1 Introduction and Motivation

Control and optimization of dynamical systems in the presence of stochastic uncertainty is a mature field with a large range of applications. A comprehensive treatment of such problems can be found in excellent books and other resources including [7, 16, 29, 68, 84, 95, 104], and [6]. To date, there exists a nearly complete theory regarding the existence and structure of optimal solutions under various formulations as well as computational methods to obtain such optimal solutions for problems with finite state and control spaces. However, there still exist substantial computational challenges involving problems with large state and action spaces, such as standard Borel spaces. For such state and action spaces, obtaining optimal policies is in general computationally infeasible.

An increasingly important avenue of research is what is often referred to as *decentralized stochastic control* or *dynamic team theory*, which involves multiple decision makers who strive for a common goal but who have access only to local information. Information structures in a decentralized control problem specify which decision maker has access to what information and the impacts of a decision maker's action on other decision maker's information. We will present a concise overview of information structures in the book, but refer the reader to more extensive resources such as [148]. Decentralized control problems may be quite challenging under certain information structures; in general few results are known regarding systematic methods to arrive at optimal solutions, and there exist problems (such as Witsenhausen's counterexample) which have defied solution attempts for more than 40 years.

Quantization, as we will demonstrate, provides a systematic constructive approach for obtaining approximately optimal solutions with guaranteed convergence properties for both classical stochastic control problems and decentralized stochastic control problems. The aim of this book is to present

N. Saldi et al., *Finite Approximations in Discrete-Time Stochastic Control*, Systems & Control: Foundations & Applications,
https://doi.org/10.1007/978-3-319-79033-6_1

our recent results regarding the approximation problem in a unified form. Our approach is to establish the convergence of approximate models obtained through quantization with a rigorous treatment involving stochastic control, probability theory, and information theory.

Additional motivation for our approach comes from networked control applications. In networked control, the transmission of real valued control actions to an actuator is not realistic when there is an information transmission constraint (physically limited by the presence of a communication channel) between a plant, a controller, or an actuator; in this case the actions of a controller must be quantized in order to be reliably transmitted to an actuator. Hence, there is a practical need to approximate control policies by policies which are computable and transmittable.

1.2 The Quantization Approach Presented in This Book

1.2.1 Markov Decision Processes and a Brief Literature Review of Approximation Methods

A discrete-time Markov decision process (MDP) is a mathematical model for sequential decision making under stochastic uncertainty that proved useful in modeling a wide range of systems in engineering, economics, and biology (see [44, 68]). An MDP can be specified by the following components: (i) the state space X and the action space A, (ii) the transition probability $p(\,\cdot\,|x,a)$ on X given $\mathsf{X} \times \mathsf{A}$ which gives the probability of the next state given the current state-action pair is (x,a), (iii) one-stage cost functions $c_t : \mathsf{X} \times \mathsf{A} \to \mathbb{R}$, $t = 0, 1, 2, \ldots$ (in general $c_t = c$ for some $c : \mathsf{X} \times \mathsf{A} \to \mathbb{R}$), and (iv) the initial distribution μ on X.

If X_t and A_t denote the state and action variables at time step t, then with these definitions, we have

$$\Pr\{X_0 \in \,\cdot\,\} = \mu(\,\cdot\,)$$

$$\Pr\{X_{t+1} \in \,\cdot\,\big|X_{\{0,t\}}, A_{\{0,t\}}\} = \Pr\{X_{t+1} \in \,\cdot\,\big|X_t, A_t\} = p(\,\cdot\,|X_t, A_t), \; t = 1, 2, \ldots$$

where $X_{\{0,t\}} = (X_0, \ldots, X_t)$ and $A_{\{0,t\}} = (A_0, \ldots, A_t)$. In this model, at each time step t, the decision maker observes the state of the system X_t and chooses an action A_t, using a decision function (control policy) π_t, depending on the observation obtained up to that time $(X_0, A_0, X_1, A_1, \ldots, A_{t-1}, X_{t-1}, X_t)$. The action can be a selection of a point from the action set, i.e., $\pi_t(X_0, A_0, X_1, A_1, \ldots, A_{t-1}, X_{t-1}, X_t) = A_t$ (deterministic policy), or a selection of a probability measure over an action set, i.e., $\pi_t(X_0, A_0, X_1, A_1, \ldots, A_{t-1}, X_{t-1}, X_t) = \Pr\{A_t \in \,\cdot\,\big|X_0, A_0, X_1, A_1, \ldots, A_{t-1}, X_{t-1}, X_t\}$ (randomized policy). The effect of choosing an action at time t is twofold: an immediate cost $c_t(X_t, A_t)$ is incurred and the state of the system evolves to a new state probabilistically according to the transition probability, i.e., $\mathsf{Pr}\{X_{t+1} \in \cdot|X_t, A_t\} = p(\cdot|X_t, A_t)$. The main objective is to choose an optimal control policy which minimizes the cost which is usually taken as one of the possibilities below. Here the expectations $\mathbb{E}[\,\cdot\,]$ are taken with respect to the probability measure on the sequence space of states and control actions induced by the initial distribution μ and control policy $\pi = \{\pi_t\}$.

(i) Discounted Cost: $\mathbb{E}\Big[\sum_{t=0}^{\infty} \beta^t c_t(X_t, A_t)\Big]$ for some $\beta \in (0, 1)$.

(ii) Average Cost: $\limsup_{T\to\infty} \frac{1}{T}\mathbb{E}\Big[\sum_{t=0}^{T-1} c_t(X_t, A_t)\Big]$.

Writing $W(\pi, x)$ to denote the cost function (either (i) or (ii)) of policy π for initial point x ($\mu = \delta_x$ point mass at x), the optimal cost, called *value function*, of the control problem is defined as

$$W^*(x) := \inf_{\pi} W(\pi, x).$$

A policy π^* is called optimal if

$$W(\pi^*, x) = W^*(x) \text{ for all } x \in \mathsf{X}.$$

In the theory of MDPs, a large body of research is devoted to studying the structural properties of optimal policies for various cost criteria. Particular attention has been paid to identifying conditions under which the optimal policy deterministically depends only on the current state. Such deterministic policies are usually called deterministic stationary policies [68] and are induced by measurable functions from the state space to the action space. Deterministic stationary policies have special significance since they constitute the smallest structured set of control policies in which one can find globally optimal policy for a large class of infinite horizon discounted cost (see, e.g., [68, 69]) or average cost optimal control problems (see, e.g., [6, 24, 91, 95]).

In the literature, various methods have been developed to tackle the approximation problem for MDPs: approximate dynamic programming, approximate value or policy iteration, approximate linear programming, simulation based techniques, neuro-dynamic programming (or reinforcement learning), state aggregation, etc. We refer the reader to [13, 17, 28, 40–42, 46, 74, 85, 98, 103, 106, 110, 134, 136, 137] and references therein for a rather complete survey of the existing techniques. Below, we summarize only a portion of these techniques.

The Approximate Value Iteration (AVI) and Approximate Policy Iteration (API) algorithms are two powerful methods to approximate an optimal (deterministic stationary) policy for an MDP (see [17, 28, 46, 126] and references therein). In AVI, the idea is to compute approximately the value function, using some prescribed basis functions, in each step of the value iteration algorithm. This way one can both approximately find the value function and construct an approximately optimal policy. In API, at each step, first an approximate cost function for a given policy is computed, again using prescribed basis functions, and then an improved policy is generated using the approximate cost function.

In the approximate linear programming approach [34, 35], a linear combination of preselected basis functions is fit to the value function of the MDP. To this end, the MDP is first formulated as a linear program over appropriate spaces. In this new formulation, by replacing functions whose optimal value corresponds to the value function with linear combination of a preselected basis functions, one can obtain a linear programming approximation. Thus, the original problem reduces to a linear program with constraints over a finite dimensional Euclidean space. One can further reduce the number of the constraints by applying constraint sampling [35]. We refer the reader to [1, 153] for applications of the approximate linear programming method to concrete problems.

Another well-known method for approximating an optimal policy is *state aggregation*. In this method, first similar states (e.g., with respect to cost and transition probabilities) are aggregated to form meta-states of a reduced MDP, and then an optimal policy is calculated for the reduced MDP (see [12, 98, 106] and references therein). An important issue regarding this method is how to efficiently aggregate states and construct a reduced MDP from the original one.

References [30, 51, 65, 134, 135] use the technique of truncating the state space when evaluating the value function in the value iteration algorithm. In these schemes, in each step the state space is truncated and the corresponding value function is calculated; the latter is proved to converge to the true value function. Then, using the truncated value function, approximately optimal policies are constructed.

With the exception of few papers, including [42, 45, 98], prior work that considered the approximation problem for MDPs in general studied either the finite horizon cost or the discounted infinite horizon cost criteria. In addition, the majority of these results are for MDPs with discrete (i.e., finite or countable) state and action spaces (see, e.g., [13, 17, 28, 46, 98, 106, 110, 134]). Those that consider general state and action spaces (see, e.g., [13, 40–43, 45, 85, 136, 137]) assume in general Lipschitz type continuity conditions on the components of the control model, sometimes along with a bounded one-stage cost function assumption. Although the assumptions imposed on the components of the control model are restrictive in the literature, the results may be very strong. For example, in [42], a simulation based approximation technique is developed leading to probabilistic guarantees on the approximation error. This paper considers Borel state and action spaces with a possibly unbounded one-stage cost function and developed a sequence of approximations using the empirical distributions generated by a probability measure ψ with respect to which the transition probability of the MDP is absolutely continuous. By imposing Lipschitz type continuity conditions on the components of the control model, the authors obtain a concentration inequality type upper bound on the accuracy of the approximation based on the Wasserstein distance of order 1 between the probability measure ψ and its empirical estimate. In [41], the authors consider the approximation problem for constrained Markov decision processes with discounted cost, where the finite model is obtained via quantization of the probability measure ψ, and develop approximation results analogous to [42] by using a linear programming formulation of the problem. In [45], the authors consider finite linear programming approximations of constrained Markov decision processes with compact state-action space for both the discounted cost and average cost criteria. They provide explicit error bounds for the approximation of the value function under a Lipschitz type regularity assumption on the transition probability and the one-stage cost function.

In partially observed MDP theory, the existence of optimal policies is usually established via converting the original partially observed stochastic control problem to a fully observed one on the belief space. However, computing an optimal policy for this fully observed model, and so for the original POMDP, using well-known dynamic programming algorithms is still a challenging problem even if the original system has finite state and action spaces, since the state space of the fully observed model is always uncountable. Hence, one would expect that the approximation methods that consider general state and action spaces for fully observed MDPs can be applied to the fully observed reduction of the POMDP to obtain finite model approximations. However, for the fully observed reduction of a POMDP, a Lipschitz type regularity condition on the transition probability is in general prohibitive. Indeed, demonstrating even weak continuity in the state-action variables, which is arguably the most relaxed regularity condition on the transition probability, is a challenging problem as was recently demonstrated in [49] for general state and action spaces (see also [27] for a control-free setup). Therefore, results developed in prior literature cannot in general be applied to computing approximately optimal policies for fully observed reduction of a POMDP, and so, for the original POMDP. In the literature there exist various results for obtaining approximately optimal policies for POMDPs (see, e.g., [89, 133, 146, 147, 156]). However, these in general assume finite state and action spaces since it is quite difficult to establish regularity conditions that are satisfied by the transition probability of the belief space MDP so that the convergence is guaranteed for general state space models. We refer the reader to the survey papers [90, 132] and the books [82, 111] for further algorithmic and computational procedures for approximating POMDPs.

1.2.2 Decentralized Stochastic Control

With the transformational advances in communications and computing technologies, systems involving multiple decision makers that have access to different information are increasingly ubiquitous. Such networked control systems constitute a particularly relevant application area for decentralized stochastic control.

Decentralized stochastic control and team decision theory have their roots in control theory and economics. Marschak [92] was perhaps the first to introduce the basic elements of teams and to provide the first steps toward the development of a *team theory*. Radner [105] provided foundational results for static teams, establishing connections between person-by-person optimality, stationarity, and team-optimality. Witsenhausen's contributions [139–143] to dynamic teams and characterization of information structures have been crucial in the progress of our understanding of dynamic teams. Detailed discussions on the design of information structures in the context of team theory and economics applications are given in [8] and [128], in addition to a rich collection of other contributions not listed here.

There are still fundamental open problems in decentralized stochastic control: Even establishing the existence and structure of optimal policies is challenging as the tools available for single decision-maker stochastic control problems are in general not applicable to decentralized systems.

The existence of optimal policies for static and a class of sequential dynamic teams has been studied recently in [60]. More specific setups and nonexistence results have been studied in [138, 144]. For a class of teams which are *convex*, one can reduce the search space to a smaller parametric class of policies (see [37, 80, 105], and for a comprehensive review, see [148]). In team decision theory, relatively few results are available on approximating optimal policies. We can only refer to [9, 52–54, 86, 87, 94] and a few references therein. With the exception of [52–54], these works in general study a specific setup (Witsenhausen's counterexample) and are mostly experimental, and as such, they do not rigorously prove the convergence of approximate solutions. In [52, 54], a class of static team problems are considered and the existence of smooth optimal strategies is studied. Under fairly strong assumptions, the existence of an optimal strategy with Lipschitz continuous partial derivatives up to some order is proved. By using this result, an error bound on the accuracy of near optimal solutions is established, where near optimal strategies are expressed as linear combinations of basis functions with adjustable parameters. In [53], the same authors investigated the approximation problem for Witsenhausen's counterexample, which does not satisfy the conditions in [52, 54], and derived an analogous error bound on the accuracy of the near optimal solutions. For the result in [53] both the error bound and the near optimal solutions depend on the knowledge of the optimal strategy for Witsenhausen's counterexample. Moreover, the method devised in [53] implicitly corresponds to the discretization of only the action spaces of the agents. Therefore, it involves only the approximation with regard to the action space, and does not correspond to a tractable approximation for the set of policies/strategies.

In team decision theory, particular attention has been paid to Witsenhausen's counterexample. This problem has puzzled the control community for more than 40 years with its philosophical impact demonstrating the challenges that arise due to a nonclassical information structure and the formidable difficulty in obtaining an optimal or suboptimal solution. In fact, optimal policies and their value are still unknown, even though the existence of an optimal policy has been proved using various methods [60, 138, 144]. Some relevant results on obtaining approximate solutions can be found in [9, 10, 53, 86, 87, 93, 94]. Lower bounds, that are not tight, building on information theoretic approaches are available in [59], see also [97]. In this book, we show that finite models obtained through the uniform quantization of the observation and action spaces lead to a sequence of policies whose cost values converge to the optimal cost. Thus, one could guarantee that an ε-optimal policy can be constructed for any $\varepsilon > 0$.

1.2.3 What This Book Presents: A Unified Quantization Approach Under Relaxed Regularity Conditions

Our aim in this book is to present, in a unified form, our recent results on the approximation of centralized and decentralized stochastic control problems with uncountable state, measurement, and action spaces through quantization methods. The coverage is extensive in that various optimality criteria and setups such as discounted cost and average cost problems, as well as cost-constrained models, partially observed problems, and decentralized stochastic control models are studied. For all of these setups, the book demonstrates that quantization provides a constructive method which is essentially independent of the system under consideration modulo certain continuity conditions, through the reduction of a system with Borel spaces to one with finite state, measurement, and action spaces. In addition to this unified view, we will also build on the powerful techniques from probability theory, source coding, and information theory to obtain bounds on the approximation performance.

As noted earlier, quantization based methods have been frequently used in the literature and in practice. However, prior results on approximation methods typically require very stringent regularity properties, such as Lipschitz continuity conditions on the transition kernel. Compared with much of the existing literature, the analysis in the book requires quite relaxed regularity conditions, such as the weak continuity of the controlled stochastic kernel. Also, the results in this book are rather general in that whenever one can establish an existence result (for an optimal control policy under discounted, constrained, or average cost criteria, or in a partially observed setup), a corresponding approximation result follows, typically with no additional assumptions. Such relaxed conditions allow for explicit constructions throughout the book. The flexibility that arises due to the relaxed regularity conditions becomes particularly evident in the context of POMDPs, where weak continuity conditions are quite natural whereas stronger conditions are typically impractical. When our goal is to obtain rates of convergence results, however, we will impose additional assumptions in our analyses.

Furthermore, unlike some of the existing results in the literature, the approach in the book may be viewed to be constructive in that the finite MDP models obtained through quantization lead to solutions that are implementable in the original setup, whose limit (as the approximating state and action sets grow unbounded) cost value coincide with the optimal cost of the true MDP.

In our approach, we consider both the information transmission point of view (i.e., discretization of actions) and the computational point of view (i.e., discretization of *states* and actions). The priority is to establish the convergence of the approximation models to the original model, even though we also provide more refined convergence rate analyses for the approximation error under further technical conditions. As noted, when the goal is to obtain only asymptotic convergence, the assumptions we impose on the control model are in general much weaker than the assumptions used for a guaranteed rate of convergence analysis, as is often the case in the prior literature.

In Part II of the book, multi-agent versions of the approximation problems are investigated in the context of decentralized stochastic control (or dynamic team theory). The results and constructions here are analogous to the single decision maker case, but the analysis techniques are quite different since the tools developed for discrete-time MDPs are in general not applicable for decentralized stochastic control problems. In this field, the operation of quantization has typically been used to show that under information structure constraints a nonlinear team policy can perform better than an optimal linear policy. This is even true for linear systems driven by Gaussian noise under quadratic cost criterion (LQG problems), as for the celebrated counterexample of Witsenhausen [10, 86] and other interesting models such as the Gaussian relay channel problem [88, 152]. Our results show that not only may quantized policies perform better than linear policies for a large class of problems, but they are actually almost optimal. The analysis in the book, therefore, rigorously certifies the near optimality of many of the numerical attempts in the literature under quite general conditions for Witsenhausen's counterexample and beyond, and paves the way for various numerical and learning methods for this emerging field.

1.3 Organization of the Book

The book consists of two parts.

Part I involves classical stochastic control problems, with a single decision maker who acts repeatedly over time with her information set growing at each time stage. In Chapter 2, we introduce the stochastic control problems considered in the subsequent chapters: we define Markov decision process, partially observed Markov decision process, constrained Markov decision process, and review some basic results. Optimality operators characterizing the optimality of deterministic stationary policies in Markov decision theory are also introduced in this chapter.

In Chapter 3, we present results regarding the finite-action approximation of stationary policies for a discrete-time Markov decision process under strong and weak continuity assumptions on the transition probability. We introduce a new family of policies, called deterministic stationary quantizer policies, and show that such policies can approximate optimal deterministic stationary policies with arbitrary precision, thus demonstrating that one can search for near optimal policies within the class of quantized control policies. We also derive explicit bounds on the approximation error in terms of the rate of the approximating quantizers. The results of Chapter 3 are based on the papers [112–114, 118].

Chapter 4 studies the finite-state approximation of discrete-time Markov decision processes. Under certain continuity conditions on the one-stage cost function and the transition probability, it is shown that a stationary policy obtained from the finite model, which is constructed by quantizing the state space of the original system on a finite grid, can approximate the optimal stationary policy with arbitrary precision. For compact-state MDPs, we obtain explicit rate of convergence bounds quantifying how the approximation improves as the number of the grid points used to quantize the state space increases. Using information theoretic arguments, the order optimality of the obtained rates of convergence rates is also established for a large class of problems. The results of Chapter 4 are based on the papers [115, 116, 119].

In Chapter 5, we consider finite model approximations of discrete-time partially observed Markov decision processes (POMDPs) under the discounted cost criterion. After converting the original partially observed stochastic control problem to a fully observed one on the belief space, the finite models are obtained through the uniform quantization of the state and action spaces of the belief space Markov decision process (MDP). Under mild assumptions on the components of the original model, it is shown that the policies obtained from these finite models are nearly optimal for the belief space MDP, and so, for the original partially observed problem. The assumptions essentially require that the belief space MDP satisfies a mild weak continuity condition. We provide examples and introduce explicit approximation procedures for the quantization of the set of probability measures on the state space of the POMDP (i.e., the belief space).

Chapter 6 studies the finite-state approximation of a discrete-time constrained Markov decision process with compact state space, under the discounted cost and average cost criteria. Using the linear programming formulation of the constrained discounted problem, we prove the convergence of the optimal value function of the finite-state model to the optimal value function of the original model. Under further continuity condition on the transition probability of the original model, we also derive a method to compute approximately optimal policies. For the average cost criterion, we also establish analogous results under drift and minorization conditions which guarantee the geometric ergodicity of Markov chains induced by stationary policies. However, in the average cost case, we use a direct method to show the convergence of the finite models instead of finite linear programming approximation. The results of Chapter 6 build, in part, on [117].

Part II of the book focuses on decentralized stochastic control problems and their applications. In Chapter 7, preliminaries on decentralized stochastic control are given. In Chapter 8, we present our results on the finite model approximation of a multi-agent stochastic control problem (team decision problem). The strategies obtained from finite models are shown to approximate the optimal cost with arbitrary precision under mild technical assumptions. In particular, we show that quantized team policies are asymptotically optimal. In Chapter 9, the results are applied to Witsenhausen's counterexample and the Gaussian relay channel problem.

For Witsenhausen's counterexample, our approximation approach provides, to our knowledge, the first rigorously established result that one can construct an ε-optimal strategy for any $\varepsilon > 0$ through a solution of a simpler problem. The results of Part II build, in part, on [120].

1.4 Notation and Conventions

The following notation will be used throughout the book. For a metric space E, we let $\mathscr{B}(\mathsf{E})$ denote the Borel σ-algebra on E. Let $\mathscr{M}(\mathsf{E})$, $\mathscr{M}_+(\mathsf{E})$, and $\mathscr{P}(\mathsf{E})$ respectively denote the set of all finite signed, finite positive, and probability measures on $(\mathsf{E}, \mathscr{B}(\mathsf{E}))$. It will be tacitly assumed that any metric space is equipped with its Borel σ-algebra and all measures on such spaces will be Borel measures [23, Definition 7.1.1]. For any $\nu \in \mathscr{M}(\mathsf{E})$ and measurable real function g on E, we define $\nu(g) := \int g d\nu$. The symbol $\delta_e \in \mathscr{P}(\mathsf{E})$ denotes the point mass at e, i.e., $\delta_e(A) = 1$ if $e \in A$ and $\delta_e(A) = 0$ if $e \notin A$ for any Borel set $A \subset \mathsf{E}$. Any Cartesian product of metric spaces will be equipped with the product Borel σ-algebra. Unless otherwise specified, the term "measurable" will refer to Borel measurability. We always equip $\mathscr{P}(\mathsf{E})$ with the Borel σ-algebra $\mathscr{B}(\mathscr{P}(\mathsf{E}))$ generated by the topology of weak convergence [20]. If E is a Borel space (i.e., Borel subset of complete and separable metric space), then $\mathscr{P}(\mathsf{E})$ is metrizable with the Prokhorov metric which makes $\mathscr{P}(\mathsf{E})$ into a Borel space [100]. We let $B(\mathsf{E})$ denote the set of all bounded measurable real functions on E and let $C_b(\mathsf{E})$ denote the set of all bounded continuous real valued functions on E. For any $u \in C_b(\mathsf{E})$ or $u \in B(\mathsf{E})$, let $\|u\| := \sup_{e \in \mathsf{E}} |u(e)|$, which turns $C_b(\mathsf{E})$ and $B(\mathsf{E})$ into Banach spaces. A sequence $\{\mu_n\}$ of measures on E is said to converge setwise to a measure μ if $\mu_n(g) \to \mu(g)$ for all $g \in B(\mathsf{E})$, or equivalently, if $\mu_n(D) \to \mu(D)$ for all $D \in \mathscr{B}(\mathsf{E})$. Given a measurable function $w : \mathsf{E} \to [1, \infty)$ and any real valued measurable function u on E, we define the w-norm of u as

$$\|u\|_w := \sup_{e \in \mathsf{E}} \frac{|u(e)|}{w(e)}.$$

We let $B_w(\mathsf{E})$ denote the set of all real valued measurable functions on E with finite w-norm and the $C_w(\mathsf{E})$ denote the set of all real valued continuous functions on E with finite w-norm. For any finite signed measure $\vartheta \in \mathscr{M}(\mathsf{E})$, we define the w-norm of ϑ as

$$\|\vartheta\|_w := \sup_{\|g\|_w \leq 1} \left| \int_{\mathsf{E}} g(e) \vartheta(de) \right|,$$

and let $\mathscr{M}_w(\mathsf{E})$ denote the set of all finite signed measures on E with finite w-norm and define $\mathscr{M}_w^+(\mathsf{E}) := \mathscr{M}_w(\mathsf{E}) \cap \mathscr{M}_+(\mathsf{E})$. When $w = 1$, $\|\vartheta\|_w$ reduces to the total variation norm, in which case we use the notation $\|\vartheta\|_{TV}$ instead of $\|\vartheta\|_w$. Let

$\mathsf{E}^n = \prod_{i=1}^{n} \mathsf{E}_i$ $(2 \leq n \leq \infty)$ be a finite or countably infinite product space. By an abuse of notation, any function g on $\prod_{j=i_1}^{i_k} \mathsf{E}_j$, where $\{i_1, \ldots, i_k\} \subseteq \{1, \ldots, n\}$, is also treated as a function on E^n by identifying it with its natural extension to E^n. The notation $V \sim \nu$ means that random variable V has distribution ν.

Part I
Finite Model Approximations in Stochastic Control

Chapter 2
Prelude to Part I

Part I involves classical stochastic control problems, with a single decision maker acting repeatedly over time with its information set growing at each time stage.

In Chapter 3, we present results regarding the finite-action approximation of stationary policies for discrete-time Markov decision processes. Chapter 4 studies the finite-state approximation of Markov decision processes. In Chapter 5, we consider finite-model approximation of partially observed Markov decision processes, and Chapter 6 studies the finite-state approximation of constrained Markov decision processes.

In this chapter, we introduce the stochastic control problems considered in the subsequent chapters; i.e., we define Markov decision processes, partially observed Markov decision processes, and constrained Markov decision processes, and review some basic results. This chapter provides some preliminary background on optimization of such controlled Markov chains. Optimality operators characterizing the optimality of stationary policies in Markov decision processes are also introduced in this chapter. A comprehensive treatment can be found in, e.g., [15, 29, 68, 84, 95], and [6].

2.1 Markov Decision Processes

A discrete-time Markov decision process (MDP) can be described by a five-tuple

$$\big(\mathsf{X}, \mathsf{A}, \{\mathsf{A}(x) : x \in \mathsf{X}\}, p, c\big),$$

where Borel spaces X and A denote the *state* and *action* spaces, respectively. The collection $\{\mathsf{A}(x) : x \in \mathsf{X}\}$ is a family of nonempty subsets of A, which form the admissible actions for the state $x \in \mathsf{X}$. The *stochastic kernel* $p(\,\cdot\,|x, a)$ denotes the *transition probability* of the next state given that previous state-action pair is (x, a)

N. Saldi et al., *Finite Approximations in Discrete-Time Stochastic Control*, Systems & Control: Foundations & Applications,
https://doi.org/10.1007/978-3-319-79033-6_2

[68]. Hence, it satisfies: (i) $p(\cdot\,|x,a)$ is an element of $\mathscr{P}(\mathsf{X})$ for all (x,a), and (ii) $p(D|\cdot,\cdot)$ is a measurable function from $\mathsf{X}\times\mathsf{A}$ to $[0,1]$ for each $D\in\mathscr{B}(\mathsf{X})$. The *one-stage cost* function $c:\mathsf{X}\times\mathsf{A}\to[0,\infty)$ is measurable. In the remainder of this book, it is assumed that $\mathsf{A}(x)=\mathsf{A}$ for all $x\in\mathsf{X}$.

Define the history spaces $\mathsf{H}_0=\mathsf{X}$ and $\mathsf{H}_t=(\mathsf{X}\times\mathsf{A})^t\times\mathsf{X}$, $t=1,2,\ldots$ endowed with the product Borel σ-algebras generated by $\mathscr{B}(\mathsf{X})$ and $\mathscr{B}(\mathsf{A})$. A *policy* is a sequence $\pi=\{\pi_t\}$ of stochastic kernels on A given H_t. The set of all policies is denoted by Π. Let Φ denote the set of stochastic kernels on A given X, and let $\mathbb{F}$ denote the set of all measurable functions from X to A. A *randomized Markov* policy is a sequence $\pi=\{\pi_t\}$ of stochastic kernels on A given X. A *deterministic Markov* policy is a sequence of stochastic kernels $\pi=\{\pi_t\}$ on A given X such that $\pi_t(\cdot\,|x)=\delta_{f_t(x)}(\cdot)$ for some $f_t\in\mathbb{F}$. The set of randomized and deterministic Markov policies are denoted by RM and M, respectively. A *randomized stationary* policy is a constant sequence $\pi=\{\pi_t\}$ of stochastic kernels on A given X such that $\pi_t(\cdot\,|x)=\varphi(\cdot\,|x)$ for all t for some $\varphi\in\Phi$. A *deterministic stationary* policy is a constant sequence of stochastic kernels $\pi=\{\pi_t\}$ on A given X such that $\pi_t(\cdot\,|x)=\delta_{f(x)}(\cdot)$ for all t for some $f\in\mathbb{F}$. The set of randomized and deterministic stationary policies are identified with the sets Φ and $\mathbb{F}$, respectively.

According to the Ionescu Tulcea theorem [68], an initial distribution μ on X and a policy π define a unique probability measure P_μ^π on $\mathsf{H}_\infty=(\mathsf{X}\times\mathsf{A})^\infty$, which is called a *strategic measure* [47]. For any μ and any policy π we can think of the MDP as a stochastic process $\{X_t,A_t\}_{t\geq 0}$ defined on the probability space $\big(\mathsf{H}_\infty,\mathscr{B}(\mathsf{H}_\infty),P_\mu^\pi\big)$, where the X_t are X-valued random variables, the A_t are A-valued random variables, and they satisfy for all $t\geq 1$

$$P_\mu^\pi(X_0\in\cdot\,)=\mu(\,\cdot\,)$$

$$P_\mu^\pi(X_t\in\cdot\,|X_{\{0,t-1\}},A_{\{0,t-1\}})=P_\mu^\pi(X_t\in\cdot\,|X_{t-1},A_{t-1})=p(\cdot\,|X_{t-1},A_{t-1})$$

$$P_\mu^\pi(A_t\in\cdot\,|X_{\{0,t\}},A_{\{0,t-1\}})=\pi_t(\cdot\,|X_{\{0,t\}},A_{\{0,t-1\}}).$$

In the sequel, we will occasionally use the following notation. For a policy π and initial distribution μ, we let $\lambda_t^{\pi,\mu}$, $\lambda_{(t)}^{\pi,\mu}$ and $\gamma_t^{\pi,\mu}$ denote the distribution of X_t, $(X_0,\ldots,X_t)$ and (X_t,A_t), respectively, i.e.,

$$X_t\sim\lambda_t^{\pi,\mu}$$

$$(X_0,\ldots,X_t)\sim\lambda_{(t)}^{\pi,\mu}$$

$$(X_t,A_t)\sim\gamma_t^{\pi,\mu}.$$

If $\mu=\delta_x$, we replace μ with x in $\lambda_t^{\pi,\mu}$, $\lambda_{(t)}^{\pi,\mu}$, and $\gamma_t^{\pi,\mu}$.

The expectation with respect to P_μ^π is denoted by $\mathbb{E}_\mu^\pi$. If $\mu=\delta_x$, we write P_x^π and $\mathbb{E}_x^\pi$ instead of $P_{\delta_x}^\pi$ and $\mathbb{E}_{\delta_x}^\pi$. The cost functions to be minimized are the β-discounted cost and the average cost, respectively, given by

$$J(\pi, x) = \mathbb{E}_x^\pi\left[\sum_{t=0}^{\infty} \beta^t c(X_t, A_t)\right],$$

$$V(\pi, x) = \limsup_{T\to\infty} \frac{1}{T}\mathbb{E}_x^\pi\left[\sum_{t=0}^{T-1} c(X_t, A_t)\right].$$

Remark 2.1 We note that the path-wise infinite sum $\sum_{t=0}^{\infty} \beta^t c(x_t, a_t)$ may not be well defined in the definition of J if c is only assumed to be measurable. However, further assumptions that will be imposed in later sections ensure that J is a well-defined function.

With this notation, the discounted and average cost value functions of the control problem are defined as

$$J^*(x) := \inf_{\pi\in\Pi} J(\pi, x),$$

$$V^*(x) := \inf_{\pi\in\Pi} V(\pi, x).$$

A policy π^* is said to be optimal if $J(\pi^*, x) = J^*(x)$ (or $V(\pi^*, x) = V^*(x)$ for the average cost) for all $x \in \mathsf{X}$. Under fairly mild conditions the set $\mathbb{F}$ of deterministic stationary policies contains optimal policies for discounted and average costs (see [49] for problems with weakly continuous transition probabilities and [68] for strongly continuous transition probabilities).

The optimality of deterministic stationary policies is usually characterized by equalities or inequalities involving optimality operators. For any real-valued function u on X, define the optimality operator T_β for $\beta \in (0, 1]$ as

$$T_\beta u(x) := \min_{a\in\mathsf{A}}\left[c(x, a) + \beta \int_{\mathsf{X}} u(y)p(dy|x, a)\right]. \tag{2.1}$$

In the literature T_β is called the *Bellman optimality operator*.

Using T_β, the so-called discounted cost optimality equation (DCOE) for $\beta \in (0, 1)$ is given by

$$J^*(x) = T_\beta J^*(x) \text{ for all } x \in \mathsf{X}; \tag{2.2}$$

that is, the value function J^* of the β-discounted cost is a fixed point of the optimality operator T_β. This fixed point equation, if the right-hand side is well defined (i.e., measurable), always holds as a result of the principle of dynamic programming. Under mild conditions (e.g. boundedness of c), a stationary policy $f^* : \mathsf{X} \to \mathsf{A}$ is discounted cost optimal if it attains the minimum in (2.2), i.e.,

$$T_\beta J^*(x) = c(x, f^*(x)) + \beta \int_{\mathsf{X}} J^*(y)p(dy|x, f^*(x)).$$

Under mild technical conditions, an optimal stationary policy f^* exists and attains the minimum in DCOE.

For the average cost criterion, the average cost optimality equation (ACOE) and inequality (ACOI) are respectively given by

$$\rho + h(x) = T_1 h(x) \tag{2.3}$$

$$\rho + h(x) \geq T_1 h(x), \tag{2.4}$$

where ρ is a real number and h is a real-valued function on X. A stationary policy f^* is average cost optimal if and only if it attains the minimum in either (2.3) or (2.4):

$$\rho + h(x) = c(x, f^*(x)) + \int_{\mathsf{X}} h(y) p(dy|x, f^*(x))$$

$$\rho + h(x) \geq c(x, f^*(x)) + \int_{\mathsf{X}} h(y) p(dy|x, f^*(x))$$

provided that $\lim_{t\to\infty} \mathbb{E}_x^{\pi}\big[h(X_t)\big]/t = 0$ for all π and $x \in \mathsf{X}$. In this case we have $V^*(x) = \rho$ for all $x \in \mathsf{X}$; that is, the average cost value function is equal to ρ for all initial points x [68, Chapter 5]. In the literature, the triplet (ρ, h, f^*) is called a *canonical triplet*. The existence of ρ and h satisfying ACOE or ACOI is in general established by the so-called "vanishing discount" approach, in which the limiting behavior (as $\beta \to 1$) of the discounted value function is used to show the existence of ρ and h [68, Chapter 5].

2.2 Partially Observed Markov Decision Processes

A discrete-time partially observed Markov decision process (POMDP) has the following components: (i) the state space X, action space A, and observation space Y, all Borel spaces, (ii) $p(\,\cdot\,|x, a)$ is the transition probability of the next state given the current state-action pair is (x, a), (iii) $r(\,\cdot\,|x)$ is the observation channel giving the probability of the current observation given the current state variable x, and (iv) the one-stage cost function $c : \mathsf{X} \times \mathsf{A} \to [0, \infty)$.

The key difference between a POMDP and an MDP, introduced in previous section, is that in a POMDP the state X_t of the system cannot be observed directly by the controller. Instead, the noisy version Y_t of the state is available through an observation channel $r(\,\cdot\,|X_t)$. These models in general arise when there is an uncertainty in measurements of the states or there are some states which cannot be measured.

To complete the description of the partially observed control model, we must specify how the controller designs its control law at each time step. To this end, define the history spaces $\mathsf{H}_0 = \mathsf{Y}$ and $\mathsf{H}_t = (\mathsf{Y} \times \mathsf{A})^t \times \mathsf{Y}$, $t = 1, 2, \ldots$

endowed with their product Borel σ-algebras generated by $\mathcal{B}(\mathsf{Y})$ and $\mathcal{B}(\mathsf{A})$. A *policy* $\pi = \{\pi_t\}$ is a sequence of stochastic kernels on A given H_t. We denote by Π the set of all policies. For any initial distribution μ and policy π we can think of the POMDP as a stochastic process $\{X_t, Y_t, A_t\}_{t\geq 0}$ defined on the probability space $(\Omega, \mathcal{B}(\Omega), P_\mu^\pi)$, where $\Omega = \mathsf{H}_\infty \times \mathsf{X}^\infty$, the X_t are X-valued random variables, the Y_t are Y-valued random variables, the A_t are A-valued random variables, and they satisfy for all $t \geq 1$

$$P_\mu^\pi(X_0 \in \cdot\,) = \mu(\,\cdot\,)$$

$$P_\mu^\pi(X_t \in \cdot\,|X_{\{0,t-1\}}, Y_{\{0,t-1\}}, A_{\{0,t-1\}}) = P_\mu^\pi(X_t \in \cdot\,|X_{t-1}, A_{t-1}) = p(\,\cdot\,|X_{t-1}, A_{t-1})$$

$$P_\mu^\pi(Y_t \in \cdot\,|X_{\{0,t\}}, Y_{\{0,t-1\}}, A_{\{0,t-1\}}) = P_\mu^\pi(Y_t \in \cdot\,|X_t) = r(\,\cdot\,|X_t)$$

$$P_\mu^\pi(A_t \in \cdot\,|X_{\{0,t\}}, Y_{\{0,t\}}, A_{\{0,t-1\}}) = \pi_t(\,\cdot\,|Y_{\{0,t\}}, A_{\{0,t-1\}})$$

where $Y_{\{0,t\}} = (Y_0, \ldots, Y_t)$. We denote by $J(\pi, \mu)$ and $V(\pi, \mu)$ the discounted and average cost functions of the policy $\pi \in \Pi$ with initial distribution μ, respectively.

In POMDPs, since the information available to the decision maker is a noisy version of the state, one cannot apply the dynamic programming principle directly as the one-stage cost function depends on the exact state information. A canonical way to overcome this difficulty is converting the original partially observed control problem to a fully observed one on the so-called belief space. Indeed, let us define

$$Z_t(\,\cdot\,) := \mathsf{Pr}\{X_t \in \cdot\,|Y_0, \ldots, Y_t, A_0, \ldots, A_{t-1}\} \in \mathcal{P}(\mathsf{X}).$$

Here, Z_t are the posterior state distributions or "beliefs" of the observer at time t in the original problem. One can prove that for any $t \geq 0$ we have

$$\mathsf{Pr}\{Z_{t+1} \in \cdot\,|Z_0, \ldots, Z_t, A_0, \ldots, A_t\} = \mathsf{Pr}\{Z_{t+1} \in \cdot\,|Z_t, A_t\} = \eta(\,\cdot\,|Z_t, A_t),$$

where η is a fixed stochastic kernel on $\mathcal{P}(\mathsf{X})$ given $\mathcal{P}(\mathsf{X}) \times \mathsf{A}$ (see Chapter 5 for the construction of η). Note that $\mathsf{Pr}\{Z_0 \in \cdot\,\} = \delta_\mu(\,\cdot\,)$. Furthermore, define the one-stage cost function $\tilde{c} : \mathcal{P}(\mathsf{X}) \times \mathsf{A} \to [0, \infty)$ as

$$\tilde{c}(z, a) := \int_{\mathsf{X}} c(x, a) z(dx).$$

Hence, we obtain a fully observed Markov decision process with the components

$$\big(\mathcal{P}(\mathsf{X}), \mathsf{A}, \eta, \tilde{c}\big).$$

This MDP is called the belief-MDP and it is equivalent to the original POMDP in the sense that for any optimal policy for the belief-MDP, one can construct a policy for the original POMDP which is optimal. The following observations are crucial in obtaining the equivalence of these two models: (i) any history vector of the belief-MDP is a function of the history vector of the original POMDP and (ii) history vectors of the belief-MDP is a sufficient statistic for the original POMDP.

Therefore, results developed for MDPs can be applied to the belief-MDP and so, to the POMDP. However, one should keep in mind that the correspondence between policies of the belief-MDP and the POMDP is quite complicated as one has to compute the so-called *nonlinear filtering equation* at each time step. Moreover, stationary policies in the belief-MDP can be history dependent in the original POMDP. Therefore, establishing structural properties of the optimal policies of the belief-MDP does not in general give much information about optimal policies of the original POMDP. However, we will see in Chapter 5 that the belief-MDP formulation of the POMDP is useful in finite-model approximation problem.

2.3 Constrained Markov Decision Processes

In some stochastic control problems constraints are introduced in the form of discounted or average costs. In this book, we also discuss these types of problems, where the cost and the constraints are given as follows. For each policy $\pi \in \Pi$ and $\beta \in (0, 1)$ define the β-discounted cost functions

$$J(\pi, \mu) = (1 - \beta)\mathbb{E}_{\mu}^{\pi}\biggl[\sum_{t=0}^{\infty} \beta^t c(X_t, A_t)\biggr],$$

$$J_l(\pi, \mu) = (1 - \beta)\mathbb{E}_{\mu}^{\pi}\biggl[\sum_{t=0}^{\infty} \beta^t d_l(X_t, A_t)\biggr] \text{ for } l = 1, \ldots, q,$$

where the $d_l, l = 1, \ldots, q$, are cost functions that place constraints on the optimization problem. Here, we normalize the usual discounted cost with the coefficient $(1 - \beta)$ to simplify some technical details. Similarly, for each policy $\pi \in \Pi$, define the average cost functions

$$V(\pi, \mu) = \limsup_{T \to \infty} \frac{1}{T}\mathbb{E}_{\mu}^{\pi}\biggl[\sum_{t=0}^{T-1} c(X_t, A_t)\biggr],$$

$$V_l(\pi, \mu) = \limsup_{T \to \infty} \frac{1}{T}\mathbb{E}_{\mu}^{\pi}\biggl[\sum_{t=0}^{T-1} d_l(X_t, A_t)\biggr] \text{ for } l = 1, \ldots, q.$$

Using above notation, the constrained MDP optimization problem can be defined as follows:

$$
\begin{aligned}
&\textbf{(CP)}\ \text{minimize}\ W(\pi, \gamma)\\
&\qquad \text{subject to}\ W_l(\pi, \gamma) \leq k_l\ (l = 1, \ldots, q),\ \pi \in \Pi,
\end{aligned}
$$

where $W \in \{J, V\}$.

For such problems, the dynamic programming principle is in general not applicable. As we will discuss in detail in *Chapter* 6, a commonly adopted method is the linear programming approach and a dual formulation, see, e.g., [67]. However, it is in general not possible to obtain constructive approximation results when one works with the dual formulation. Further discussion on these aspects will be presented later. We refer the reader to [67] and [3] for further details on constrained MDPs.

Chapter 3
Finite-Action Approximation of Markov Decision Processes

3.1 Introduction

In this chapter, we study the finite-action approximation of optimal control policies for discrete-time Markov decision processes (MDPs) with Borel state and action spaces, under discounted and average cost criteria. One main motivation for considering this problem stems from the optimal information transmission problem in networked control systems. In many applications of networked control, perfect transmission of the control actions to an actuator is infeasible when there is a communication channel of finite capacity between a controller and an actuator. Hence, the actions of the controller must be discretized (quantized) to facilitate reliable transmission. Although the problem of optimal information transmission from a plant/sensor to a controller has been studied extensively (see, e.g., [148] and references therein), much less is known about the problem of transmitting actions from a controller to an actuator. Such transmission schemes usually require a simple encoding/decoding rule since the actuator does not have the computational capability of the controller to use complex algorithms. For this reason, time-invariant scalar quantization is a practically useful encoding method for controller-actuator communication.

The finite-action approximation problem will be treated for two cases: (i) MDPs with strongly continuous transition probability and (ii) MDPs with weakly continuous transition probability. Under further assumptions, we also obtain explicit performance bounds on the approximation error in terms of the cardinality of the finite action set.

In the approximation result under the weak continuity assumption, we approximate the optimality operators associated with the dynamic programming equations characterizing optimality of deterministic stationary policies, which is a standard technique in approximate dynamic programming. However, for the strong continuity case we develop a novel technique in which we approximate strategic measures

N. Saldi et al., *Finite Approximations in Discrete-Time Stochastic Control*, Systems & Control: Foundations & Applications,
https://doi.org/10.1007/978-3-319-79033-6_3

induced by policies on the infinite product of state and action spaces. One advantage of this technique is that it can be used to study the approximation problem for constrained Markov decision processes for which the dynamic programming principle no longer hold.

3.2 Quantizing the Action Space

We first give the definition of a quantizer from the state to the action space.

Definition 3.1 *A measurable function* $q : \mathsf{X} \to \mathsf{A}$ *is called a* quantizer *from* X *to* A *if the range of* q*, i.e.,* $q(\mathsf{X}) = \{q(x) \in \mathsf{A} : x \in \mathsf{X}\}$*, is finite.*

The elements of $q(\mathsf{X})$ (the possible values of q) are called the *levels* of q. The rate $R = \log_2 |q(\mathsf{X})|$ of a quantizer q (approximately) represents the number of bits needed to losslessly encode the output levels of q using binary codewords of equal length. Let $\mathcal{Q}$ denote the set of all quantizers from X to A. A *deterministic stationary quantizer* policy is a constant sequence $\pi = \{\pi_t\}$ of stochastic kernels on A given X such that $\pi_t(\,\cdot\,|x) = \delta_{q(x)}(\,\cdot\,)$ for all t for some $q \in \mathcal{Q}$. For any finite set $\Lambda \subset \mathsf{A}$, let $\mathcal{Q}(\Lambda)$ denote the set of all elements in $\mathcal{Q}$ having range Λ. Analogous with $\mathbb{F}$, the set of all deterministic stationary quantizer policies induced by $\mathcal{Q}(\Lambda)$ will be identified with the set $\mathcal{Q}(\Lambda)$.

Our main objective is to show that under some mild conditions there exists a sequence of finite subsets $\Lambda_n \subset \mathsf{A}$ such that value functions of the MDP$_n$ with components $\{\mathsf{X}, \Lambda_n, p, c\}$ converge to the value function of the original MDP as $n \to \infty$. More formally, the sequence of finite subsets $\{\Lambda_n\}_{n\geq 1}$ must satisfy the following.

(P) If the set of deterministic stationary policies $\mathbb{F}$ is optimal for the MDP, then for the discounted cost

$$\lim_{n\to\infty} \inf_{q\in\mathcal{Q}(\Lambda_n)} J(q, x) = J^*(x) \text{ for all } x \in \mathsf{X}$$

or for the average cost

$$\lim_{n\to\infty} \inf_{q\in\mathcal{Q}(\Lambda_n)} V(q, x) = V^*(x) \text{ for all } x \in \mathsf{X}.$$

3.2.1 *Finite Action Model*

In the remainder of this chapter we assume that the action space A is compact. Let d_{A} denote the metric on A. Since A is assumed compact and thus totally bounded, there exists a sequence of finite sets $\Lambda_n = \{a_{n,1}, \ldots, a_{n,k_n}\} \subset \mathsf{A}$ such that for each n,

$$\min_{i \in \{1,\ldots,k_n\}} d_{\mathsf{A}}(a, a_{n,i}) < 1/n \text{ for all } a \in \mathsf{A}.$$

In other words, Λ_n is a $1/n$-net in A. In the rest of this chapter, the sequence $\{\Lambda_n\}_{n \geq 1}$ is used to solve problem **(P)**. To simplify the notation, we define the mapping $\Upsilon_n : \mathbb{F} \to \mathscr{Q}(\Lambda_n)$ as

$$\Upsilon_n(f)(x) = \arg\min_{a \in \Lambda_n} d_{\mathsf{A}}(f(x), a), \tag{3.1}$$

where ties are broken so that $\Upsilon_n(f)(x)$ is measurable. Hence, for all $f \in \mathbb{F}$, we have

$$\sup_{x \in \mathsf{X}} d_{\mathsf{A}}\big(f(x), \Upsilon_n(f)(x)\big) < 1/n; \tag{3.2}$$

that is $\Upsilon_n(f)$ converges *uniformly* to f as $n \to \infty$.

3.3 Near Optimality of Quantized Policies Under Strong Continuity

In this section we consider problem **(P)** for MDPs with strongly continuous transition probability. Note that the transition probability p is said to be strongly continuous if it is setwise continuous in the action variable. We impose the assumptions below on the components of the Markov decision process; additional assumptions will be made for the average cost problem in Section 3.3.2.

Assumption 3.1

(a) The one-stage cost function c is bounded and $c(x, \cdot)$ is continuous for all $x \in \mathsf{X}$.
(b) The stochastic kernel $p(\cdot\,|x, a)$ is setwise continuous in $a \in \mathsf{A}$, i.e., if $a_k \to a$, then $p(\cdot\,|x, a_k) \to p(\cdot\,|x, a)$ setwise for all $x \in \mathsf{X}$.
(c) A is compact.

Remark 3.2 Note that if X is countable, then $B(\mathsf{X}) = C_b(\mathsf{X})$ (here X is endowed with the discrete topology) which implies the equivalence of setwise convergence and weak convergence. Hence, the results developed in this section are applicable to MDPs having weakly continuous (in the action variable) transition probabilities when the state space is countable.

Remark 3.3 Note that any MDP can be modeled by a discrete-time dynamical system of the form $X_{t+1} = F(X_t, A_t, V_t)$, where $\{V_t\}$ is a sequence of independent and identically distributed (i.i.d.) random variables with values in some space V and common distribution ν. In many applications, the function F has a well-behaved structure and is in the form $F(x, a, v) = H(x, a)G(v)$ or $F(x, a, v) = H(x, a) + G(v)$, e.g., the *fisheries management model* [68, p. 5], the *cash balance model*

[48], and the *Pacific halibut fisheries management model* [42]. In these systems, Assumption 3.1-(b) holds for common noise processes. For instance, if ν admits a continuous density, which is often the case in practice, then Assumption 3.1-(b) usually holds. We refer the reader to [48, Section 4] for a discussion on the relevance of the setwise continuity assumption on inventory control problems. In addition, the widely studied and practically important case of the additive noise system in our Example 3.7 in the next section also satisfies Assumption 3.1-(b).

3.3.1 Discounted Cost

In this section we suppose that Assumption 3.1 holds. Recall that for any $\nu \in \mathscr{P}(\mathsf{E})$ and measurable real function g on E, we define $\nu(g) := \int g d\nu$. Since the one-stage cost function c is bounded, the discounted cost satisfies the following (recall the notation introduced in Section 2.1)

$$\sup_{f \in \mathbb{F}} \sum_{t=T+1}^{\infty} \beta^t \gamma_t^{f,x}(c) \to 0 \text{ as } T \to \infty. \tag{3.3}$$

We now define the ws^{∞} topology on $\mathscr{P}(\mathsf{H}_{\infty})$ which was first introduced by Schäl in [121]. Let $\mathscr{C}(\mathsf{H}_0) = B(\mathsf{X})$ and let $\mathscr{C}(\mathsf{H}_t)$ $(t \geq 1)$ be the set of real valued functions g on H_t such that $g \in B(\mathsf{H}_t)$ and $g(x_0, \cdot, x_1, \cdot, \ldots, x_{t-1}, \cdot, x_t) \in C_b(\mathsf{A}^t)$ for all $(x_0, \ldots, x_t) \in \mathsf{X}^{t+1}$. The ws^{∞} topology on $\mathscr{P}(\mathsf{H}_{\infty})$ is defined as the smallest topology which renders all mappings $P \mapsto P(g)$, $g \in \bigcup_{t=0}^{\infty} \mathscr{C}(\mathsf{H}_t)$, continuous.

The following theorem is the main result of this section. It states that for any $f \in \mathbb{F}$, the discounted cost function of policy $\Upsilon_n(f) \in \mathscr{Q}(\Lambda_n)$ converges to the discounted cost function of policy f as $n \to \infty$, which implies that the discounted cost value function of the MDP$_n$ converges to the discounted cost value function of the original MDP.

Theorem 3.4 *Let $f \in \mathbb{F}$ and let $\{\Upsilon_n(f)\}$ be the quantized approximations of f. Then, $J(\Upsilon_n(f), x) \to J(f, x)$ as $n \to \infty$.*

The proof of Theorem 3.4 requires the following proposition which is proved in Section 3.6.

Proposition 3.5 *For any $f \in \mathbb{F}$ and all $x \in \mathsf{X}$, the strategic measures $\{P_x^{\Upsilon_n(f)}\}$ induced by the quantized approximations $\{\Upsilon_n(f)\}$ converge to the strategic measure P_x^f of f in the ws^{∞} topology.*

Proof of Theorem 3.4. Let $q_n = \Upsilon_n(f)$. Proposition 3.5 implies that $\gamma_t^{q_n,x}(c) \to \gamma_t^{f,x}(c)$ as $n \to \infty$ for all t. Then, we have

$$\limsup_{n\to\infty} |J(q_n, x) - J(f, x)| \leq \limsup_{n\to\infty} \sum_{t=0}^{\infty} |\beta^t \gamma_t^{q_n,x}(c) - \beta^t \gamma_t^{f,x}(c)|$$

$$\leq \lim_{n\to\infty} \sum_{t=0}^{T} |\beta^t \gamma_t^{q_n,x}(c) - \beta^t \gamma_t^{f,x}(c)| + 2 \sup_{f'\in\mathbb{F}} \sum_{t=T+1}^{\infty} \beta^t \gamma_t^{f',x}(c).$$

Since the first and second terms in the last expression converge to zero as $T \to \infty$ by Proposition 3.5 and (3.3), respectively, the proof is complete. □

Remark 3.6 Notice that this proof implicitly shows that J is sequentially continuous in the ws^{∞} topology with respect to the strategic measures.

The following is a generic example frequently considered in the theory of Markov decision processes (see [71]).

Example 3.7 Let us consider an additive-noise system given by

$$X_{t+1} = F(X_t, A_t) + V_t, \; t = 0, 1, 2, \ldots$$

where $\mathsf{X} = \mathbb{R}^m$ and the $\{V_t\}$ is a sequence of independent and identically distributed (i.i.d.) $\mathbb{R}^m$-valued random vectors whose common distribution has a continuous probability density function. A nondegenerate Gaussian distribution satisfies this condition. Let the action space A be a compact subset of $\mathbb{R}^d$ for some $d \geq 1$ and assume the one-stage cost function c satisfies Assumption 3.1-(a), and that $F(x, \cdot)$ is continuous for all $x \in \mathsf{X}$. It is straightforward to show that Assumption 3.1-(b) holds under these conditions. Hence, Theorem 3.4 holds for this system.

3.3.2 *Average Cost*

In contrast to the discounted cost criterion, the expected average cost is in general not sequentially continuous for the ws^{∞} topology with respect to strategic measures under practical assumptions. Hence, in this section we develop an approach based on the convergence of the sequence of invariant probability measures of the state process $\{X_t\}_{t=0}^{\infty}$ under quantized stationary policies to solve **(P)** for the average cost criterion.

Observe that any deterministic stationary policy f defines a stochastic kernel on X given X via

$$Q_f(\cdot|x) := \lambda_1^{f,x}(\cdot) = p(\cdot|x, f(x)). \tag{3.4}$$

Let us write $Q_f g(x) := \lambda_1^{f,x}(g)$. If Q_f admits an ergodic invariant probability measure ν_f, then by [70, Theorem 2.3.4 and Proposition 2.4.2], there exists an invariant set with full ν_f measure such that for all x in that set we have

$$V(f,x) = \limsup_{T\to\infty} \frac{1}{T} \sum_{t=0}^{T-1} \gamma_t^{f,x}(c)$$
$$= \lim_{T\to\infty} \frac{1}{T} \sum_{t=0}^{T-1} \lambda_t^{f,x}(c_f) = \nu_f(c_f), \tag{3.5}$$

where $c_f(x) := c(x, f(x))$. Let $\mathsf{M}_f \in \mathscr{B}(\mathsf{X})$ be the set of all $x \in \mathsf{X}$ such that the convergence in (3.5) holds. Hence, $\nu_f(\mathsf{M}_f) = 1$ if ν_f exists. The following assumptions will be imposed in this section.

Assumption 3.2 *Suppose Assumption 3.1 holds and in addition we have*

(e) *For any $f \in \mathbb{F}$, Q_f has a unique invariant probability measure ν_f.*
(f1) *The set of invariant measures $\Gamma_{\mathbb{F}} := \{\nu \in \mathscr{P}(\mathsf{X}) : \nu Q_f = \nu \text{ for some } f \in \mathbb{F}\}$ is relatively sequentially compact in the setwise topology.*
(f2) *There exists $x \in \mathsf{X}$ such that for all $B \in \mathscr{B}(\mathsf{X})$, $\lambda_t^{f,x}(B) \to \nu_f(B)$ uniformly in $f \in \mathbb{F}$.*
(g) $\mathsf{M} := \bigcap_{f\in\mathbb{F}} \mathsf{M}_f \neq \emptyset$.

Remark 3.8 If we assume that the process under any stationary policy is positive Harris recurrent [96], then $\mathsf{M}_f = \mathsf{X}$ for all f.

Before we state the main result of this section, we will derive conditions under which the conditions in Assumption 3.2 hold. To begin with, Assumption 3.2-(e),(f2),(g) are satisfied if the transition probability Q_f is *geometrically ergodic* for any $f \in \mathbb{F}$; that is, for any $f \in \mathbb{F}$, there exists a probability measure ν_f such that for all $x \in \mathsf{X}$ and for all n

$$\|Q_f^n(\,\cdot\,|x) - \nu_f\|_w \leq Rw(x)\kappa^n, \tag{3.6}$$

where $R > 0$, $\kappa \in (0,1)$, and $w : \mathsf{X} \to [1,\infty)$. Geometric ergodicity condition (3.6) with $w = 1$ is true under any of the conditions Ri, $i \in \{0, 1, 1(a), 1(b), 2, \ldots, 5\}$ in [72]; that is,

(R0) There exist a state $x^* \in \mathsf{X}$ and a number $\alpha > 0$ such that

$$p(\{x^*\}|x, f(x)) \geq \alpha \ \text{ for all } x \in \mathsf{X} \text{ and } \ f \in \mathbb{F}.$$

(R1) There exist a positive integer n and a nontrivial measure ν on X such that

$$p^n(\,\cdot\,|x, f(x)) \geq \nu(\,\cdot\,) \ \text{ for all } x \in \mathsf{X} \text{ and } f \in \mathbb{F}.$$

(R1a) For each $f \in \mathbb{F}$, the transition probability $p(dy|x, f(x))$ has a density $\rho(y|x, f(x))$ with respect to a σ-finite measure m on X, and there exist $\varepsilon > 0$ and $C \in \mathscr{B}(\mathsf{X})$ such that $m(C) > 0$ and

$$\rho(y|x, f(x)) \geq \varepsilon \text{ for all } y \in C, x \in \mathsf{X}, \text{ and } f \in \mathbb{F}.$$

(R1b) For each $f \in \mathbb{F}$, the transition probability $p(dy|x, f(x))$ has a density $\rho(y|x, f(x))$ with respect to a σ-finite measure m on X, and $\rho(y|x, f(x)) \geq \rho_0(y)$ for all $x, y \in \mathsf{X}$, $f \in \mathbb{F}$, where ρ_0 is a nonnegative measurable function with $m(\rho_0) > 0$.

(R2) There exist a positive integer n and a measure ν on X such that

$$\nu(\mathsf{X}) < 2 \text{ and } p^n(\,\cdot\,|x, f(x)) \leq \nu(\,\cdot\,) \text{ for all } x \in \mathsf{X} \text{ and } f \in \mathbb{F}.$$

(R3) There exist a positive integer n and a positive number $\alpha < 1$ such that

$$\|p^n(\,\cdot\,|x, f(x)) - p^n(\,\cdot\,|y, \tilde{f}(y))\|_{TV} \leq 2\alpha \text{ for all } x, y \in \mathsf{X} \text{ and } f, \tilde{f} \in \mathbb{F}.$$

(R4) There exist a positive integer n and a positive number $\alpha < 1$ such that

$$\|p^n(\,\cdot\,|x, f(x)) - p^n(\,\cdot\,|y, f(y))\|_{TV} \leq 2\alpha \text{ for all } x, y \in \mathsf{X} \text{ and } f \in \mathbb{F}.$$

(R5) There exist a positive integer n and a positive number α for which the following holds: For each $f \in \mathbb{F}$, there is a probability measure ν_f on X such that

$$p^n(\,\cdot\,|x, f(x)) \geq \alpha \nu_f \text{ for all } x \in \mathsf{X}.$$

Moreover, we have $\mathsf{M} = \mathsf{X}$ in Assumption 3.2-(g) if at least one of the conditions Ri, $i \in \{0, 1, 1(a), 1(b), 2, \ldots, 6\}$ holds. We refer to [96, Theorem 16.02] and [69, Section 7.5.D] for further conditions that imply geometric ergodicity. In these references, "drift" and "minorization" type conditions are in general used to obtain geometric ergodicity. A variant of these conditions will be presented for the finite-state approximation problem in Section 4.2.2 and Section 4.3.2.

The next step is to find sufficient conditions for Assumption 3.2-(e),(f1),(g) to hold. Observe that the stochastic kernel p on X given $\mathsf{X} \times \mathsf{A}$ can be written as a measurable mapping from $\mathsf{X} \times \mathsf{A}$ to $\mathcal{P}(\mathsf{X})$ if $\mathcal{P}(\mathsf{X})$ is equipped with its Borel σ-algebra generated by the weak topology [70], i.e., $p(\,\cdot\,|x, a) : \mathsf{X} \times \mathsf{A} \to \mathcal{P}(\mathsf{X})$. We impose the following assumption:

(e1) There exists a finite positive measure ζ on X such that $p(\,\cdot\,|x, a) \leq \zeta(\,\cdot\,)$ for all $x \in \mathsf{X}$ and $a \in \mathsf{A}$.

Proposition 3.9 *Suppose (e1) holds. Then, for any $f \in \mathbb{F}$, Q_f has an invariant probability measure ν_f. Furthermore, $\Gamma_{\mathbb{F}}$ is sequentially relatively compact in the setwise topology. Hence, (e1) implies Assumption 3.2-(f1). In addition, if these invariant measures are unique, then Assumption 3.2-(e),(g) also hold with $\mathsf{M} = \mathsf{X}$ in Assumption 3.2-(g).*

Proof For any $f \in \mathbb{F}$ and $x \in \mathsf{X}$, define

$$Q_{f,x}^{(T)}(\cdot) := \frac{1}{T} \sum_{t=0}^{T-1} \lambda_t^{f,x}(\cdot).$$

Clearly, $Q_{f,x}^{(T)} \leq \zeta$ for all T. Hence, by [70, Corollary 1.4.5] there exists a subsequence $\{Q_{f,x}^{(T_k)}\}$ which converges setwise to some probability measure ν_f. Following the same steps in [63, Theorem 4.17] one can show that $\nu_f(g) = \nu_f(Q_f g)$, for all $g \in B(\mathsf{X})$. Hence, ν_f is an invariant probability measure for Q_f.

Furthermore, (e1) implies that $\nu_f \leq \zeta$ for all $\nu_f \in \Gamma_{\mathbb{F}}$. Thus, $\Gamma_{\mathbb{F}}$ is relatively sequentially compact in the setwise topology by again [70, Corollary 1.4.5].

Finally, for any f, if the invariant measure ν_f is unique, then every setwise convergent subsequence of the relatively sequentially compact sequence $\{Q_{f,x}^{(T)}\}$ must converge to ν_f. Hence, $Q_{f,x}^{(T)} \to \nu_f$ setwise which implies that $V(f,x) = \lim_{T\to\infty} Q_{f,x}^{(T)}(c_f) = \nu_f(c_f)$ for all $x \in \mathsf{X}$ since $c_f(x) := c(x, f(x)) \in B(\mathsf{X})$. Thus, $\mathsf{M} = \mathsf{X}$ in Assumption 3.2-(g). □

The following theorem is the main result of this section. It states that for any $f \in \mathbb{F}$, the average cost function of $\Upsilon_n(f) \in \mathcal{Q}(\Lambda_n)$ converges to the average cost function of f as $n \to \infty$. In other words, the average cost value function of MDP_n converges to the average cost value function of the original MDP.

Theorem 3.10 *Let $x \in \mathsf{M}$. Then, we have $V(\Upsilon_n(f), x) \to V(f,x)$ as $n \to \infty$, under Assumption 3.2 with (f1) or (f2).*

Proof Let $\Upsilon_n(f) = q_n$. Let Q_f and Q_{q_n} be the stochastic kernels, respectively, for f and q_n defined in (3.4). By Assumption 3.2-(e), Q_f and Q_{q_n} have unique, and so ergodic, invariant probability measures ν_f and ν_{q_n}, respectively. Since $x \in \mathsf{M}$, we have $V(q_n, x) = \nu_{q_n}(c_{q_n})$ and $V(f,x) = \nu_f(c_f)$, where $c_{q_n}(x) = c(x, q_n(x))$ and $c_f(x) = c(x, f(x))$. Observe that $c_{q_n}(x) \to c_f(x)$ for all x by Assumption 3.1-(a). Hence, if we prove that $\nu_{q_n} \to \nu_f$ setwise, then by [122, Theorem 2.4] we have $V(q_n, x) \to V(f,x)$. We prove this first under (f1) and then under (f2).

I) Proof under assumption (f1)

We show that every setwise convergent subsequence $\{\nu_{q_{n_l}}\}$ of $\{\nu_{q_n}\}$ must converge to ν_f. Then, since $\Gamma_{\mathbb{F}}$ is relatively sequentially compact in the setwise topology, there is at least one setwise convergent subsequence $\{\nu_{q_{n_l}}\}$ of $\{\nu_{q_n}\}$, which implies the result.

Let $\nu_{q_{n_l}} \to \nu$ setwise for some $\nu \in \mathcal{P}(\mathsf{X})$. We will show that $\nu = \nu_f$, or equivalently, ν is an invariant probability measure for Q_f. For simplicity, we write $\{\nu_{q_l}\}$ instead of $\{\nu_{q_{n_l}}\}$. Let $g \in B(\mathsf{X})$. By Assumption 3.2-(e) we have $\nu_{q_l}(g) = \nu_{q_l}(Q_{q_l} g)$. Since $Q_{q_l} g(x) \to Q_f g(x)$ for all x by Assumption 3.1-(b) and $\nu_{q_l} \to \nu$ setwise, we have $\nu_{q_l}(Q_{q_l} g) \to \nu(Q_f g)$ by [122, Theorem 2.4]. On the other hand, since $\nu_{q_l} \to \nu$ setwise we have $\nu_{q_l}(g) \to \nu(g)$. Thus $\nu(g) = \nu(Q_f g)$. Since g is arbitrary, ν is an invariant probability measure for Q_f.

II) Proof under assumption (f2)

Observe that for all t and $x \in \mathsf{X}$, $\lambda_t^{q_n,x} \to \lambda_t^{f,x}$ setwise as $n \to \infty$ since $P_x^{q_n} \to P_x^f$ in the ws^∞ topology (see Proposition 3.5). Let $B \in \mathscr{B}(\mathsf{X})$ be given and fix some $\varepsilon > 0$. By Assumption 3.2-(f2) we can choose T large enough such that $|\lambda_T^{\tilde{f},x}(B) - \nu_{\tilde{f}}(B)| < \varepsilon/3$ for all $\tilde{f} \in \{f, q_1, q_2, \cdots\}$. For this T, choose N large enough such that $|\lambda_T^{q_n,x}(B) - \lambda_T^{f,x}(B)| < \varepsilon/3$ for all $n \geq N$. Thus, for all $n \geq N$ we have $|\nu_{q_n}(B) - \nu_f(B)| \leq |\nu_{q_n}(B) - \lambda_T^{q_n,x}(B)| + |\lambda_T^{q_n,x}(B) - \lambda_T^{f,x}(B)| + |\lambda_T^{f,x}(B) - \nu_f(B)| < \varepsilon$. Since ε is arbitrary, we obtain $\nu_{q_n}(B) \to \nu_f(B)$, which completes the proof. □

Example 3.11 Let us consider the additive-noise system in Example 3.7 with the same assumptions. Furthermore, we assume F is bounded and the density of the noise process is positive everywhere. Observe that for any $f \in \mathbb{F}$, if Q_f has an invariant probability measure, then it has to be unique [70, Lemma 2.2.3] since there cannot exist disjoint invariant sets due to the positivity of the density function of the noise process. Since this system satisfies (e1) and R1(a) due to the boundedness of F, Assumption 3.2-(e),(f1),(f2),(g) hold with $\mathsf{M} = \mathsf{X}$. This means that Theorem 3.10 holds for an additive noise system under the above conditions.

3.4 Near Optimality of Quantized Policies Under Weak Continuity

Requiring that the transition probability of the system be weakly continuous in state-action variables (i.e., weak continuity) is a fairly mild assumption compared to the setwise continuity in the action variable (i.e., strong continuity). Indeed, the latter condition is even prohibitive for certain stochastic systems such as fully observed reduction of a POMDP (see Example 5.1). In this section we consider the additive noise model in order to more explicitly highlight this.

Example 3.12 Consider an additive-noise system given in Example 3.7. In this system, the continuity of F in (x, a) (which holds for most practical systems) is sufficient to imply the weak continuity of the transition probability, and no assumptions are needed on the noise process (not even the existence of a density is required). Hence, weak continuity does not restrict the noise model and is satisfied by almost all systems in the applications, whereas other conditions, such as strong continuity or continuity in total variation distance, hold only if the noise is well behaved in addition to the continuity of F. Recall from Example 3.7 that for setwise continuity, it is required that F is continuous in a for every x, the noise admits a density, and this density is continuous.

3.4.1 Discounted Cost

The following assumptions will be imposed for the discounted cost case. These assumptions are commonly used in the literature for studying discounted Markov decision processes with unbounded one-stage cost and weakly continuous transition probability.

Assumption 3.3

(a) The one-stage cost function c is continuous.
(b) The stochastic kernel $p(\cdot|x,a)$ is weakly continuous in $(x,a) \in \mathsf{X} \times \mathsf{A}$, i.e., if $(x_k,a_k) \to (x,a)$, then $p(\cdot|x_k,a_k) \to p(\cdot|x,a)$ weakly.
(c) A is compact.
(d) There exist nonnegative real numbers M and $\alpha \in [1, \frac{1}{\beta})$, and a continuous weight function $w : \mathsf{X} \to [1,\infty)$ such that for each $x \in \mathsf{X}$, we have

$$\sup_{a\in\mathsf{A}} c(x,a) \leq Mw(x), \tag{3.7}$$

$$\sup_{a\in\mathsf{A}} \int_{\mathsf{X}} w(y)p(dy|x,a) \leq \alpha w(x), \tag{3.8}$$

and $\int_{\mathsf{X}} w(y)p(dy|x,a)$ is continuous in (x,a).

Recall the Bellman optimality operator T_β defined in (2.1). Since β is fixed here, we will write T instead of T_β. For any real-valued measurable function u on X, $Tu : \mathsf{X} \to \mathbb{R}$ is given by

$$Tu(x) := \min_{a\in\mathsf{A}} \biggl[c(x,a) + \beta \int_{\mathsf{X}} u(y)p(dy|x,a)\biggr]. \tag{3.9}$$

Recall also that $C_w(\mathsf{X})$ and $B_w(\mathsf{X})$ denote the set of all real valued continuous and measurable functions on X with finite w-norm, respectively.

Lemma 3.13 *For any $u \in C_w(\mathsf{X})$, the function $l_u(x,a) := \int_{\mathsf{X}} u(y)p(dy|x,a)$ is continuous in (x,a).*

Proof For any nonnegative continuous function g on X, the function $l_g(x,a) = \int_{\mathsf{X}} g(y)p(dy|x,a)$ is lower semi-continuous in (x,a) if p is weakly continuous (see, e.g., [68, Proposition E.2]). Define the nonnegative continuous function g by letting $g = bw + u$, where $b = \|u\|_w$. Then l_g is lower semi-continuous. Since $l_u = l_g - bl_w$ and l_w is continuous by Assumption 3.3-(d), l_u is lower semi-continuous. Analogously, define the nonnegative continuous function v by letting $v = -u + bw$. Then l_v is lower semi-continuous. Since $l_u = bl_w - l_v$ and l_w is continuous by Assumption 3.3-(d), l_u is also upper semi-continuous. Therefore, l_u is continuous. □

Lemma 3.14 *Let* Y *be any of the compact sets* A *or* Λ_n, $n \geq 1$. *Define the operator* T_{Y} *on* $B_w(\mathsf{X})$ *by letting*

$$T_{\mathsf{Y}} u(x) := \min_{a \in \mathsf{Y}} \left[c(x,a) + \beta \int_{\mathsf{X}} u(y) p(dy|x,a) \right].$$

Then T_{Y} *maps* $C_w(\mathsf{X})$ *into itself. Moreover,* $C_w(\mathsf{X})$ *is closed with respect to the* w-*norm.*

Proof Note that $T_{\mathsf{Y}} u(x) = \min_{a \in \mathsf{Y}} \big(c(x,a) + \beta l_u(x,a) \big)$, where l_u is continuous by Lemma 3.13, and therefore, $T_{\mathsf{Y}} u$ is also continuous by [16, Proposition 7.32]. Since T_{Y} maps $B_w(\mathsf{X})$ into itself, $T_{\mathsf{Y}} u \in C_w(\mathsf{X})$.

For the second statement, let $u_n \in C_w(\mathsf{X})$ converge to u in w-norm. It is sufficient to prove that u is continuous. Let $x_k \to x_0$. Since $B := \{x_0, x_1, x_2, \ldots\}$ is compact, w is bounded on B. Therefore, $u_n \to u$ uniformly on B, which implies that $\lim_{k \to \infty} u(x_k) = u(x_0)$. This completes the proof. □

Lemma 3.14 implies that T (defined in (3.9)) maps $C_w(\mathsf{X})$ into itself. It can also be proved that T is a contraction operator with modulus $\sigma := \beta\alpha$ (see [69, Lemma 8.5.5]); that is,

$$\|Tu - Tv\|_w \leq \sigma \|u - v\|_w \text{ for all } u, v \in C_w(\mathsf{X}).$$

The following theorem is a known result in the theory of Markov decision processes (see e.g., [69, Section 8.5, p.65]).

Theorem 3.15 *Suppose Assumption 3.3 holds. Then the discounted cost value function* J^* *is the unique fixed point in* $C_w(\mathsf{X})$ *of the contraction operator* T, *i.e.,* $J^* = TJ^*$. *Furthermore, a deterministic stationary policy* f^* *is optimal if and only if*

$$J^*(x) = c(x, f^*(x)) + \beta \int_{\mathsf{X}} J^*(y) p(dy|x, f^*(x)). \tag{3.10}$$

Finally, there exists a deterministic stationary policy f^* *which is optimal, so it satisfies (3.10).*

Using the definition in Lemma 3.14, for $n \geq 1$, let $T_n := T_{\Lambda_n}$ (this operator will be used to approximate $T = T_{\mathsf{A}}$). Note that T_n is the Bellman optimality operator for MDP$_n$ having components $\big\{\mathsf{X}, \Lambda_n, p, c\big\}$. Analogous with T, it can be shown that T_n is a contraction operator with modulus $\sigma = \alpha\beta$ mapping $C_w(\mathsf{X})$ into itself. Let $J_n^* \in C_w(\mathsf{X})$ (the discounted cost value function of MDP$_n$) denote the fixed point of T_n.

The following theorem is the main result of this section which states that the discounted cost value function of MDP$_n$ converges to the discounted cost value function of the original MDP.

Theorem 3.16 *For any compact set $K \subset \mathsf{X}$ we have*

$$\lim_{n\to\infty} \sup_{x\in K} |J_n^*(x) - J^*(x)| = 0.$$

Therefore, $\lim_{n\to\infty} |J_n^*(x) - J^*(x)| = 0$ *for all* $x \in \mathsf{X}$.

To prove Theorem 3.16, we need the following auxiliary results whose proofs will be given in Section 3.6.

Lemma 3.17 *For any compact subset K of X and for any $\varepsilon > 0$, there exists a compact subset K_ε of X such that*

$$\sup_{(x,a)\in K\times\mathsf{A}} \int_{K_\varepsilon^c} w(y)p(dy|x,a) < \varepsilon.$$

Lemma 3.18 *Let $\{u_n\}$ be a sequence in $C_w(\mathsf{X})$ with $\sup_n \|u_n\|_w := L < \infty$. If u_n converges to $u \in C_w(\mathsf{X})$ uniformly on each compact subset of X, then for any $f \in \mathbb{F}$ and compact subset K of X we have*

$$\lim_{n\to\infty} \sup_{x\in K} \left| \int_{\mathsf{X}} u_n(y)p(dy|x, f_n(x)) - \int_{\mathsf{X}} u(y)p(dy|x, f(x)) \right| = 0,$$

where $f_n = \Upsilon_n(f)$.

Let us define $v^0 = v_n^0 = 0$, and $v^{t+1} = Tv^t$ and $v_n^{t+1} = T_n v_n^t$ for $t \geq 1$; that is, $\{v^t\}_{t\geq 1}$ and $\{v_n^t\}_{t\geq 1}$ are successive approximations to the discounted cost value functions of the original MDP and MDP$_n$, respectively. Lemma 3.14 implies that v^t and v_n^t are in $C_w(\mathsf{X})$ for all t and n. By [69, Theorem 8.3.6, p. 47], [69, (8.3.34), p. 52], and [69, Section 8.5, p. 65] we have

$$v^t(x) \leq J^*(x) \leq M\frac{w(x)}{1-\sigma}, \tag{3.11}$$

$$\|v^t - J^*\|_w \leq M\frac{\sigma^t}{1-\sigma}, \tag{3.12}$$

and

$$v_n^t(x) \leq J_n^*(x) \leq M\frac{w(x)}{1-\sigma}, \tag{3.13}$$

$$\|v_n^t - J_n^*\|_w \leq M\frac{\sigma^t}{1-\sigma}. \tag{3.14}$$

Since for each n and u, $Tu \leq T_n u$, we also have $v^t \leq v_n^t$ for all $t \geq 1$ and $J^* \leq J_n^*$.

Lemma 3.19 *For any compact set $K \subset \mathsf{X}$ and $t \geq 1$, we have*

$$\lim_{n\to\infty} \sup_{x\in K} |v_n^t(x) - v^t(x)| = 0. \tag{3.15}$$

Proof We prove (3.15) by induction. For $t = 1$, the claim holds since $v^0 = v_n^0 = 0$ and c is uniformly continuous on $K \times \mathsf{A}$ for any compact subset K of X. Assume the claim is true for $t \geq 1$. We fix a compact set $K \subset \mathsf{X}$. Let f_t^* denote the selector of $Tv^t = v^{t+1}$; that is,

$$v^{t+1}(x) = Tv^t(x) = c(x, f_t^*(x)) + \beta \int_{\mathsf{X}} v^t(y) p(dy|x, f_t^*(x)),$$

and let $f_{t,n}^* := \Upsilon_n(f_t^*)$ (see (3.1)). By (3.11) and (3.13) we have

$$v^t(x) \leq M \frac{w(x)}{1-\sigma} \tag{3.16}$$

$$v_n^t(x) \leq M \frac{w(x)}{1-\sigma}, \tag{3.17}$$

for all t and n. For each $n \geq 1$, we have

$$\begin{aligned}
&\sup_{x\in K} \big(v_n^{t+1}(x) - v^{t+1}(x)\big) \ \ (\text{since } v^{t+1} \leq v_n^{t+1}) \\
&= \sup_{x\in K} \bigg(\min_{\Lambda_n} \Big[c(x,a) + \beta \int_{\mathsf{X}} v_n^t(y) p(dy|x,a) \Big] - \min_{\mathsf{A}} \Big[c(x,a) + \beta \int_{\mathsf{X}} v^t(y) p(dy|x,a) \Big] \bigg) \\
&\leq \sup_{x\in K} \bigg(\Big[c(x, f_{t,n}^*(x)) + \beta \int_{\mathsf{X}} v_n^t(y) p(dy|x, f_{t,n}^*(x)) \Big] \\
&\quad - \Big[c(x, f_t^*(x)) + \beta \int_{\mathsf{X}} v^t(y) p(dy|x, f_t^*(x)) \Big] \bigg) \\
&\leq \sup_{x\in K} \big| c(x, f_{t,n}^*(x)) - c(x, f_t^*(x)) \big| \\
&\quad + \beta \sup_{x\in K} \bigg| \int_{\mathsf{X}} v_n^t(y) p(dy|x, f_{t,n}^*(x)) - \int_{\mathsf{X}} v^t(y) p(dy|x, f_t^*(x)) \bigg|.
\end{aligned}$$

Note that in the last expression, the first term goes to zero as $n \to \infty$ since c is uniformly continuous on $K \times \mathsf{A}$ and $f_{t,n}^* \to f_t^*$ uniformly, and the second term goes to zero by Lemma 3.18, (3.16), and (3.17). □

Now, we can prove Theorem 3.16 with the aid of Lemma 3.19.

Proof of Theorem 3.16. Let us fix any compact set $K \subset \mathsf{X}$. Since w is bounded on K, it is enough to prove $\lim_{n\to\infty} \sup_{x\in K} \frac{|J_n^*(x) - J^*(x)|}{w(x)}$. We have

$$\sup_{x \in K} \frac{|J_n^*(x) - J^*(x)|}{w(x)} \leq \sup_{x \in K} \frac{|J_n^*(x) - v_n^t(x)|}{w(x)}$$
$$+ \sup_{x \in K} \frac{|v_n^t(x) - v^t(x)|}{w(x)} + \sup_{x \in K} \frac{|v^t(x) - J^*(x)|}{w(x)}$$
$$\leq 2M \frac{\sigma^t}{1-\sigma} + \sup_{x \in K} \frac{|v_n^t(x) - v^t(x)|}{w(x)} \quad \text{(by (3.12) and (3.14))}.$$

Since $w \geq 1$, $\sup_{x \in K} \frac{|v_n^t(x) - v^t(x)|}{w(x)} \to 0$ as $n \to \infty$ for all t by Lemma 3.19. Hence, the last expression can be made arbitrarily small. Since $t \geq 1$ is arbitrary and $\sigma \in (0, 1)$, this completes the proof. □

3.4.2 Average Cost

In this section we prove an approximation result analogous to Theorem 3.16 for the average cost case. To do this, some new assumptions are needed on the components of the original MDP in addition to Assumption 3.3. A version of these assumptions was used in [129] and [55] to show the existence of the solution to the Average Cost Optimality Equality (ACOE) and Inequality (ACOI).

Assumption 3.4 *Suppose Assumption 3.3 holds with (3.8) replaced by condition (e) below. Moreover, suppose there exist a nondegenerate finite measure λ on X and a positive constant b such that*

(e) $\int_{\mathsf{X}} w(y) p(dy|x, a) \leq \alpha w(x) + b$ *for all* $(x, a) \in \mathsf{X} \times \mathsf{A}$, *where* $\alpha \in (0, 1)$.
(f) $p(D|x, a) \geq \lambda(D)$ *for all* $(x, a) \in \mathsf{X} \times \mathsf{A}$ *and* $D \in \mathscr{B}(\mathsf{X})$.
(g) The weight function w is λ-integrable.

Remark 3.20 Note that in the remainder of this section, we let $b = \int w d\lambda$ without loss of generality. Indeed, (i) if $b \leq \int w d\lambda$, we can replace b with $\int w d\lambda$ and the inequality in Assumption 3.4-(e) is still true. Conversely, (ii) if $\int w d\lambda < b$, then by first increasing the value of α so that $\lambda(\mathsf{X}) + \alpha > 1$, and then adding a constant k to w, where

$$k := \frac{b - \int w d\lambda}{\lambda(\mathsf{X}) + \alpha - 1},$$

we obtain $b < \int w d\lambda$. Then, as in (i) above, we set $b = \int w d\lambda$ and Assumption 3.4-(e) now holds for the new w and α.

In this section, we suppose that Assumption 3.4 holds. The following theorem states that there is a solution to the average cost optimality equation (ACOE) and the stationary policy which minimizes this ACOE is an optimal policy.

Theorem 3.21 *There exist $f^* \in \mathbb{F}$, $h^* \in C_w(\mathsf{X})$, and ρ^* such that the triplet (h^*, f^*, ρ^*) satisfies the average cost optimality equality (ACOE) and therefore, for all $x \in \mathsf{X}$*

$$V(f^*, x) = \inf_{\pi \in \Pi} V(\pi, x) := V^*(x) = \rho^*.$$

Proof Define $\hat{p}(\,\cdot\,|x, a) := p(\,\cdot\,|x, a) - \lambda(\,\cdot\,)$ which is a nonnegative finite measure for all (x, a) by Assumption 3.4-(f). Using this, let us define the operator

$$Fu(x) := \min_{a \in \mathsf{A}} \left[c(x, a) + \int_{\mathsf{X}} u(y) \hat{p}(dy|x, a) \right].$$

By Lemma 3.13 the function inside the minimization is continuous in (x, a) if $u \in C_w(\mathsf{X})$. Then by Lemma 3.14, F maps $C_w(\mathsf{X})$ into itself.

We claim that F is a contraction operator on $C_w(\mathsf{X})$ with modulus α. Indeed, let $u, v \in C_w(\mathsf{X})$. Then we have

$$\begin{aligned}
|Fu(x) - Fv(x)| &\leq \sup_{a \in \mathsf{A}} \left| \int_{\mathsf{X}} u(y) \hat{p}(dy|x, a) - \int_{\mathsf{X}} v(y) \hat{p}(dy|x, a) \right| \\
&= \sup_{a \in \mathsf{A}} \left| \int_{\mathsf{X}} \frac{u(y)}{w(y)} w(y) \hat{p}(dy|x, a) - \int_{\mathsf{X}} \frac{u(y)}{w(y)} w(y) \hat{p}(dy|x, a) \right| \\
&\leq \|u - v\|_w \sup_{a \in \mathsf{A}} \int_{\mathsf{X}} w(y) \hat{p}(dy|x, a) \\
&\leq \|u - v\|_w \alpha w(x) \text{ (by Assumption 3.4-(e) and Remark 3.20.)}
\end{aligned}$$

Hence $\|Fu - Fv\|_w \leq \alpha \|u - w\|_w$ as claimed.

This result implies that F has a unique fixed point $h^* \in C_w(\mathsf{X})$; that is,

$$\begin{aligned}
h^*(x) &= \min_{a \in \mathsf{A}} \left[c(x, a) + \int_{\mathsf{X}} h^*(y) p(dy|x, a) - \int_{\mathsf{X}} h^*(y) \lambda(dy) \right] \\
&= T_1 h^*(x) - \int_{\mathsf{X}} h^*(y) \lambda(dy).
\end{aligned}$$

Let $f^* \in \mathbb{F}$ be a stationary policy which achieves the above minimum and set $\rho^* = \int_{\mathsf{X}} h^*(y) \lambda(dy)$. Then the triplet (h^*, f^*, ρ^*) satisfies ACOE. Since $h^* \in C_w(\mathsf{X})$, one can prove that under any policy and initial point, we have $\lim_{t \to \infty} \mathbb{E}[h^*(X_t)]/t = 0$ by Assumption 3.4-(e). Hence f^* is an optimal policy by [68, Theorem 5.2.4]. This completes the proof. □

Note that all the statements in Theorem 3.21 are also valid for MDP_n with an optimal policy f_n^* and a canonical triplet (h_n^*, f_n^*, ρ_n^*). Analogous with F, define the contraction operator F_n (with modulus α) corresponding to MDP_n as

$$F_n u(x) := \min_{a \in \Lambda_n} \left[c(x,a) + \int_{\mathsf{X}} u(y) \hat{p}(dy|x,a) \right].$$

Then $h_n^* \in C_w(\mathsf{X})$ is a fixed point of F_n.

The next theorem is the main result of this section which states that the average cost value function, denoted as V_n^*, of MDP$_n$ converges to the average cost value function V^* of the original MDP.

Theorem 3.22 *We have*

$$\lim_{n \to \infty} |V_n^* - V^*| = 0,$$

where V^* and V_n^* $(n \geq 1)$ do not depend on x.

Let us define $u^0 = u_n^0 = 0$, and $u^{t+1} = F u^t$ and $u_n^{t+1} = F_n u_n^t$ for $t \geq 1$; that is, $\{u^t\}_{t \geq 1}$ and $\{u_n^t\}_{t \geq 1}$ are successive approximations to h^* and h_n^*, respectively. Lemma 3.14 implies that u^t and u_n^t are in $C_w(\mathsf{X})$ for all t and n.

Lemma 3.23 *For all $u, v \in C_w(\mathsf{X})$ and $n \geq 1$, we have: (i) if $u \leq v$, then $Fu \leq Fv$ and $F_n u \leq F_n v$; (ii) $Fu \leq F_n u$.*

Proof The results follow from the fact that $\hat{p}(\,\cdot\,|x,a) \geq 0$. □

Lemma 3.23 implies that $u^0 \leq u^1 \leq u^2 \leq \ldots \leq h^*$ and $u_n^0 \leq u_n^1 \leq u_n^2 \leq \ldots \leq h_n^*$. Note that $\|u^1\|_w, \|u_n^1\|_w \leq M$ by Assumption 3.3-(d). Since

$$\|h^*\|_w \leq \|h^* - u^1\|_w + \|u^1\|_w = \|Fh^* - Fu^0\|_w + \|u^1\|_w \leq \alpha \|h^*\|_w + \|u^1\|_w$$

and

$$\|h_n^*\|_w \leq \|h_n^* - u_n^1\|_w + \|u_n^1\|_w = \|F_n h_n^* - F_n u_n^0\|_w + \|u_n^1\|_w \leq \alpha \|h_n^*\|_w + \|u_n^1\|_w,$$

we have

$$u^t(x) \leq h^*(x) \leq M \frac{w(x)}{1-\alpha},$$

and

$$u_n^t(x) \leq h_n^*(x) \leq M \frac{w(x)}{1-\alpha}.$$

By the inequalities above and since $\|u^t - h^*\|_w \leq \alpha^t \|h\|_w$ and $\|u_n^t - h_n^*\|_w \leq \alpha^t \|h_n\|_w$, we also have

$$\|u^t - h^*\|_w \leq M \frac{\alpha^t}{1-\alpha},$$

and

$$\|u_n^t - h_n^*\|_w \leq M \frac{\alpha^t}{1-\alpha}.$$

By Lemma 3.23, for each n and v, we have $Fv \leq F_n v$. Therefore, by the monotonicity of F and the fact $u^0 = u_n^0 = 0$, we have

$$\begin{aligned} u^t &\leq u_n^t \\ h^* &\leq h_n^*, \end{aligned} \tag{3.18}$$

for all t and n.

Lemma 3.24 *For any compact set $K \subset \mathsf{X}$ and $t \geq 1$, we have*

$$\lim_{n\to\infty} \sup_{x\in K} |u_n^t(x) - u^t(x)| = 0. \tag{3.19}$$

Proof Note that for each $t \geq 1$, by dominated convergence theorem and $\lambda(w) < \infty$, we have $\lambda(u_n^t) \to \lambda(u^t)$ if $u_n^t \to u^t$ pointwise. The rest of the proof can be done using the same arguments as in the proof of Lemma 3.19 and so we omit the details. □

Lemma 3.25 *For any compact set $K \subset \mathsf{X}$, we have*

$$\lim_{n\to\infty} \sup_{x\in K} |h_n^*(x) - h^*(x)| = 0.$$

Proof The lemma can be proved using the same arguments as in the proof of Theorem 3.16. □

We are now in a position to prove Theorem 3.22

Proof of Theorem 3.22. Recall that $V^* = \rho^*$ and $V_n^* = \rho_n^*$, and that they satisfy the following ACOEs:

$$\begin{aligned} h^*(x) + \rho^* &= \min_{a\in\mathsf{A}} \left[c(x,a) + \int_{\mathsf{X}} h^*(y) p(dy|x,a) \right] \\ &= c(x, f^*(x)) + \int_{\mathsf{X}} h^*(y) p(dy|x, f^*(x)) \\ h_n^*(x) + \rho_n^* &= \min_{a\in\Lambda_n} \left[c(x,a) + \int_{\mathsf{X}} h_n^*(y) p(dy|x,a) \right] \\ &= c(x, f_n^*(x)) + \int_{\mathsf{X}} h_n^*(y) p(dy|x, f_n^*(x)). \end{aligned}$$

Note that $h_n^* \geq h^*$ (see (3.18)) and $\rho_n^* \geq \rho^*$. For each n, let $f_n := \Upsilon_n(f^*)$. Then for any $x \in \mathsf{X}$ we have

$$\begin{aligned}
\limsup_{n\to\infty} \big(h_n^*(x) + \rho_n^*\big) &= \limsup_{n\to\infty} \bigg(\min_{a\in\Lambda_n} \bigg[c(x,a) + \int_{\mathsf{X}} h_n^*(y) p(dy|x,a) \bigg] \bigg) \\
&= \limsup_{n\to\infty} \bigg(c(x, f_n^*(x)) + \int_{\mathsf{X}} h_n^*(y) p(dy|x, f_n^*(x)) \bigg) \\
&\leq \limsup_{n\to\infty} \bigg(c(x, f_n(x)) + \int_{\mathsf{X}} h_n^*(y) p(dy|x, f_n(x)) \bigg) \\
&= c(x, f^*(x)) + \int_{\mathsf{X}} h^*(y) p(dy|x, f^*(x)) \qquad (3.20) \\
&= h^*(x) + \rho^* \\
&\leq \liminf_{n\to\infty} \big(h_n^*(x) + \rho_n^*\big),
\end{aligned}$$

where (3.20) follows from Lemma 3.18 and the fact that h_n^* converges to h^* uniformly on each compact subset K of X and $\sup_n \|h_n^*\|_w \leq \frac{M}{1-\alpha}$. Since $\lim_{n\to\infty} h_n^*(x) = h^*(x)$ by Lemma 3.25, we have $\lim_{n\to\infty} \rho_n^* = \rho^*$. This completes the proof. □

3.5 Rates of Convergence

In Sections 3.3 and 3.4 we established the convergence of the finite-action models MDP_n to the original model. In this section we derive performance bounds on the approximation errors due to quantization of the action space in terms of the number of points used to discretize action space.

Specifically we show that the approximation error in **(P)** can be explicitly controlled by the number of points used to discretize the action space. To do this we need to impose a new set of assumptions in this section.

Assumption 3.5

(h) A *is infinite compact subset of* $\mathbb{R}^d$ *for some* $d \geq 1$ *and* d_{A} *is the Euclidean distance inherited from* $\mathbb{R}^d$.

(j) c *is bounded and* $|c(x,\tilde{a}) - c(x,a)| \leq K_1 d_{\mathsf{A}}(\tilde{a},a)$ *for all* x*, and some* $K_1 \geq 0$.

(k) $\|p(\,\cdot\,|x,\tilde{a}) - p(\,\cdot\,|x,a)\|_{TV} \leq K_2 d_{\mathsf{A}}(\tilde{a},a)$ *for all* x*, and some* $K_2 \geq 0$.

(l) There exist positive constants C *and* $\kappa \in (0,1)$ *such that for all* $f \in \mathbb{F}$*, there is a (necessarily unique) probability measure* $\nu_f \in \mathscr{P}(\mathsf{X})$ *satisfying* $\|\lambda_t^{f,x} - \nu_f\|_{TV} \leq C\kappa^t$ *for all* $x \in \mathsf{X}$ *and* $t \geq 1$.

Assumption 3.5-(l) implies that for any policy $f \in \mathbb{F}$, the stochastic kernel Q_f, defined in (3.4), has a unique invariant probability measure ν_f and satisfies *geometric ergodicity* with $w = 1$ (see Section 3.3.2). Hence, Assumption 3.5-(l) holds under any of the conditions Ri, $i \in \{0, 1, 1(a), 1(b), 2, \ldots, 5\}$ introduced in Section 3.3.2. Moreover, one can explicitly compute the constants C and κ for certain systems. For instance, consider the additive-noise system in Example 3.7 with Gaussian noise. Let $\mathsf{X} = \mathbb{R}$. Assume F has the bounded range $[-L, L]$ for some $L > 0$. Then, Assumption 3.5-(l) holds with $C = 2$ and $\kappa = 1 - \varepsilon L$, where $\varepsilon = \frac{1}{\sigma\sqrt{2\pi}} \exp^{-(2L)^2/2\sigma^2}$. For further conditions that imply Assumption 3.5-(l) we refer [70, 72, 96, 109].

The following example gives sufficient conditions for the additive noise system under which Assumption 3.5-(j),(k),(l) hold.

Example 3.26 Consider the additive-noise system in Example 3.7. In addition to the assumptions there, suppose $\{F(x, \cdot) : x \in \mathsf{X}\}$ is uniformly Lipschitz and the common density g of the V_t is Lipschitz on all compact subsets of X. Note that a Gaussian density has these properties. Let $c(x, a) := \|x - a\|^2$. Under these conditions, Assumption 3.5-(j),(k) hold for the additive noise system. If we further assume that F is bounded, then Assumption 3.5-(l) holds as well.

As can be easily seen by considering rectangular grids, if A is a compact subset of $\mathbb{R}^d$ then for all $n \geq 1$ there exist a constant $\alpha > 0$ and finite subsets $\Lambda_n \subset \mathsf{A}$ with cardinality $|\Lambda_n| = n$ such that

$$\max_{x \in \mathsf{A}} \min_{y \in \Lambda_n} d_{\mathsf{A}}(x, y) \leq \alpha(1/n)^{1/d}. \tag{3.21}$$

In the remainder of this section, we replace Λ_n defined in Section 3.2.1 with Λ_n satisfying (3.21). As a result, we have the following lemma.

Lemma 3.27 *Let $\mathsf{A} \subset \mathbb{R}^d$ be compact. Then for any $f \in \mathbb{F}$ the sequence $\{\Upsilon_n(f)\}$ satisfies* $\sup_{x \in \mathsf{X}} d_{\mathsf{A}}(\Upsilon_n(f)(x), f(x)) \leq \alpha(1/n)^{1/d}$.

The following proposition is the key result in this section. It is proved in Section 3.6

Proposition 3.28 *Let $f \in \mathbb{F}$ and let $\{q_n\}$ be the sequence of quantized approximations of f, i.e., $\Upsilon_n(f) = q_n$. For any initial point x and $t \geq 1$, we have*

$$\|\lambda_t^{f,x} - \lambda_t^{q_n,x}\|_{TV} \leq \alpha K_2(2t - 1)(1/n)^{1/d} \tag{3.22}$$

if Assumption 3.5-(h),(j),(k) hold.

3.5.1 Discounted Cost

In this section we suppose that Assumption 3.5-(h),(j),(k) hold. The following result gives a rate of convergence in the approximation problem for the discounted cost criterion. Its proof essentially follows from Proposition 3.28.

Theorem 3.29 *Let $f \in \mathbb{F}$ and let $\{q_n\}$ be the sequence of quantized approximations of f. For any initial point x and all $n \geq 1$, we have*

$$|J(f, x) - J(q_n, x)| \leq K(1/n)^{1/d}, \tag{3.23}$$

where $K = \frac{\alpha}{1-\beta}(K_1 - \beta K_2 M + \frac{2\beta M K_2}{1-\beta})$ *with* $M := \|c\|$.

Proof Let $c_f(x) = c(x, f(x))$ and $c_{q_n}(x) = c(x, q_n(x))$. For any fixed n we have

$$\begin{aligned}
|J(f, x) - J(q_n, x)| &= \left| \sum_{t=0}^{\infty} \beta^t \lambda_t^{f,x}(c_f) - \sum_{t=0}^{\infty} \beta^t \lambda_t^{q_n,x}(c_{q_n}) \right| \\
&\leq \sum_{t=0}^{\infty} \beta^t \big(|\lambda_t^{f,x}(c_f) - \lambda_t^{f,x}(c_{q_n})| + |\lambda_t^{f,x}(c_{q_n}) - \lambda_t^{q_n,x}(c_{q_n})| \big) \\
&\leq \sum_{t=0}^{\infty} \beta^t \big(\|c_f - c_{q_n}\| + \|\lambda_t^{f,x} - \lambda_t^{q_n,x}\|_{TV} M \big) \\
&\leq \sum_{t=0}^{\infty} \beta^t \big(\sup_{x \in \mathsf{X}} d_{\mathsf{A}}(f(x), q_n(x)) K_1 + \|\lambda_t^{f,x} - \lambda_t^{q_n,x}\|_{TV} M \big) \\
&\leq \sum_{t=0}^{\infty} \beta^t \big((1/n)^{1/d} \alpha K_1 \big) + \sum_{t=1}^{\infty} \beta^t \big((1/n)^{1/d} (2t-1) K_2 \alpha M \big) \qquad (3.24) \\
&= (1/n)^{1/d} \frac{\alpha}{1-\beta} (K_1 - \beta K_2 M + \frac{2\beta M K_2}{1-\beta}).
\end{aligned}$$

Here (3.24) follows from Assumption 3.5-(j), Proposition 3.28, and Lemma 3.27, completing the proof. □

3.5.2 Average Cost

In this section we suppose that Assumption 3.5 holds. Note Assumption 3.5-(l) implies that for any $f \in \mathbb{F}$, $V(f, x) = \nu_f(c_f)$ for all x, where $c_f(x) = c(x, f(x))$ and ν_f is the unique invariant probability measure associated with f. The following theorem essentially follows from Proposition 3.28 and Assumption 3.5-(l). It gives a convergence rate for the average cost approximation problem.

Theorem 3.30 *For* $f \in \mathbb{F}$ *let* $\{q_n\}$ *be the sequence of quantized approximations of* f, *i.e.,* $\Upsilon_n(f) = q_n$. *Then, for any* $x \in \mathsf{X}$ *and* $t, n \geq 1$, *we have*

$$|V(f, x) - V(q_n, x)| \leq 2MC\kappa^t + K_t(1/n)^{1/d}, \tag{3.25}$$

where $K_t = \big((2t-1)K_2\alpha M + K_1\alpha\big)$ *and* $M := \|c\|$.

Proof For any n and $x \in \mathsf{X}$, we have

$$\begin{aligned}
|V(f,x) - V(q_n,x)| &= |\nu_f(c_f) - \nu_{q_n}(c_{q_n})| \leq |\nu_f(c_f) - \nu_f(c_{q_n})| + |\nu_f(c_{q_n}) - \nu_{q_n}(c_{q_n})| \\
&\leq \|c_f - c_{q_n}\| + \|\nu_f - \nu_{q_n}\|_{TV} M \\
&\leq \sup_{x \in \mathsf{X}} K_1 d_{\mathsf{A}}(f(x), q_n(x)) + \|\nu_f - \nu_{q_n}\|_{TV} M \text{ (by Assumption 3.5-(j))} \\
&\leq (1/n)^{1/d} K_1 \alpha + \big(\|\nu_f - \lambda_t^{f,x}\|_{TV} + \|\lambda_t^{f,x} - \lambda_t^{q_n,x}\|_{TV} + \|\lambda_t^{q_n,x} - \nu_{q_n}\|_{TV}\big) M \\
&\leq (1/n)^{1/d} K_1 \alpha + \big(2C\kappa^t + (1/n)^{1/d}(2t-1)K_2\alpha\big) M \qquad (3.26) \\
&= 2MC\kappa^t + \big((2t-1)K_2\alpha M + K_1\alpha\big)(1/n)^{1/d},
\end{aligned}$$

where (3.26) follows from Assumption 3.5-(l) and Proposition 3.28. □

The dependence of the upper bound in (3.25) on t can be eliminated by minimizing the upper bound with respect to t for each n. To ease the notation, define the constants $I_1 := 2MC$, $I_2 := K_1\alpha - K_2\alpha M$, and $I_3 := 2K_2\alpha M$. Then the upper bound in (3.25) can be written as

$$I_1\kappa^t + I_2(1/n)^{1/d} + I_3(1/n)^{1/d} t. \tag{3.27}$$

It is straightforward to compute that

$$t'(n) := \ln\big(\frac{n^{1/d}}{I_4}\big)\frac{1}{\ln(\frac{1}{\kappa})}$$

is the zero of the derivative of the convex term in (3.27), where $I_4 := \frac{I_3}{I_1 \ln(\frac{1}{\kappa})}$. Letting $t = \lceil t'(n) \rceil$ in (3.27), we obtain the following result.

Corollary 3.31 *For $f \in \mathbb{F}$ let $\{q_n\}$ be the sequence of quantized approximations of f, i.e., $\Upsilon_n(f) = q_n$. Then, for any $x \in \mathsf{X}$ and $n \geq 1$, we have*

$$|V(f,x) - V(q_n,x)| \leq (I_1 I_4 + I_2)(1/n)^{1/d} + \frac{I_3}{\ln(1/\kappa)}(1/n)^{1/d} \ln\big(\frac{n^{1/d}}{I_4}\big).$$

3.5.3 Order Optimality

The following example demonstrates that the order of approximation error in Theorems 3.29 and 3.30 cannot be better than $O((\frac{1}{n})^{\frac{1}{d}})$. More precisely, we exhibit a simple standard example where we can lower bound the approximation error for the optimal stationary policy by $L(1/n)^{1/d}$ for some positive constant L.

In what follows $h(\cdot)$, $h(\cdot|\cdot)$, and $I(\cdot;\cdot)$ denote differential entropy, conditional differential entropy, and mutual information, respectively [33].

Consider the additive-noise system:

$$X_{t+1} = F(X_t, A_t) + V_t,\ t = 0, 1, 2, \ldots,$$

where $\mathsf{X} = \mathsf{A} \subset \mathbb{R}^d$ and the $\{V_t\}$ is a sequence i.i.d. random vectors whose common distribution has density g supported on some compact subset V of $\mathbb{R}^d$. We assume that

$$\sup_{(x,a)\in\mathbb{R}^d\times\mathbb{R}^d} \frac{\|F(x,a)\|}{\|x\| + \|a\|} < 1/2.$$

We choose V such that $\mathsf{X} = \mathsf{A}$ can be restricted to be a compact subset of $\mathbb{R}^d$. For simplicity suppose that the initial distribution μ has the same density g. It is assumed that the differential entropy $h(g) := -\int_{\mathsf{X}} g(x)\log g(x)dx$ is finite. Let the one-stage cost function be $c(x,a) := \|x - a\|$. Clearly, the optimal stationary policy f^* is the identity $f^*(x) = x$, having optimal cost $W(f,\mu) = 0$, where $W \in \{J, W\}$. Let q_n be the nth quantized approximation of f^*. Fix any n and define $D_t := \mathbb{E}_{\mu}^{q_n}\big[c(X_t, A_t)\big]$ for all t. Note that D_t can be viewed as a distortion corresponding to the quantizer q_n applied to the source X_t. Let $R(D_t)$ be the (information) rate-distortion function [33, p. 307] of the source X_t. In the following, we will use an information theoretic argument to obtain a lower bound on the performance of all such quantizers with a given number of levels. Since $R(D_t)$ is the minimum asymptotically achievable rate for the distortion level D_t [33, p. 306], we have

$$\log(n) \geq R(D_t).$$

Then, by the Shannon lower bound (SLB) [145, p. 12] for the rate-distortion function $R(D_t)$ we have

$$\begin{aligned}
\log n &\geq R(D_t) \\
&= \inf_{\mathbb{E}[\|X_t - A_t\|]\leq D_t} I(X_t; A_t) \\
&= \inf_{\mathbb{E}[\|X_t - A_t\|]\leq D_t} \big\{h(X_t) - h(X_t|A_t)\big\} \\
&= \inf_{\mathbb{E}[\|X_t - A_t\|]\leq D_t} \big\{h(X_t) - h(X_t - A_t|A_t)\big\} \\
&\geq \inf_{\mathbb{E}[\|X_t - A_t\|]\leq D_t} \big\{h(X_t) - h(X_t - A_t)\big\} \\
&= h(X_t) - \sup_{\mathbb{E}[\|X_t - A_t\|]\leq D_t} h(X_t - A_t) \\
&\geq h(X_t) + \theta(D_t)
\end{aligned}$$

$$
\begin{aligned}
&= h(F(X_{t-1}, A_{t-1}) + V_{t-1}) + \theta(D_t) \\
&\geq h(F(X_{t-1}, A_{t-1}) + V_{t-1} | X_{t-1}, A_{t-1}) + \theta(D_t) \\
&= h(V_{t-1}) + \theta(D_t), \qquad (3.28)
\end{aligned}
$$

where

$$\theta(D_t) = -d + \log\left(\frac{1}{dV_d\Gamma(d)}\Big(\frac{d}{D_t}\Big)^d\right),$$

V_d is the volume of the unit sphere $S_d = \{x : \|x\| \leq 1\}$, and Γ is the gamma function. Here, (3.28) follows from the independence of V_{t-1} and the pair (X_{t-1}, A_{t-1}). Note that $h(V_{t-1}) = h(g)$ for all t. Hence, by inverting the function θ we obtain $D_t \geq L(1/n)^{1/d}$, where

$$L := \frac{d}{2}\left(\frac{2^{h(g)}}{dV_d\Gamma(d)}\right)^{1/d}.$$

This gives

$$|J(f^*, \mu) - J(q_n, \mu)| \geq \frac{L}{1-\beta}(1/n)^{1/d}$$

and

$$|V(f^*, \mu) - V(q_n, \mu)| \geq L(1/n)^{1/d}.$$

3.6 Proofs

Proof of Proposition 3.5

To ease the notation let $\Upsilon_n(f) = q_n$. Suppose $g \in \mathscr{C}(\mathsf{H}_t)$ for some t. Then we have $P_x^{q_n}(g) = \lambda_{(t)}^{q_n,x}(g_{q_n})$ and $P_x^f(g) = \lambda_{(t)}^{f,x}(g_f)$, where $g_{q_n} = g(x_0, q_n(x_0), \ldots, q_n(x_{t-1}), x_t)$ and $g_f = g(x_0, f(x_0), \ldots, f(x_{t-1}), x_t)$. Since g is continuous in the "a'' terms by definition and q_n converges to f, we have $g_{q_n} \to g_f$. Hence, by [122, Theorem 2.4] it is enough to prove that $\lambda_{(t)}^{q_n,x} \to \lambda_{(t)}^{f,x}$ setwise as $n \to \infty$, which we prove by induction.

Clearly, $\lambda_{(1)}^{q_n,x} \to \lambda_{(1)}^{f,x}$ setwise by Assumption 3.1-(b). Assume the claim is true for some $t \geq 1$. For any $h \in B(\mathsf{X}^{t+2})$ we can write $\lambda_{(t+1)}^{q_n,x}(h) = \lambda_{(t)}^{q_n,x}\big(\lambda_1^{q_n,x_t}(h)\big)$ and $\lambda_{(t+1)}^{f,x}(h) = \lambda_{(t)}^{f,x}\big(\lambda_1^{f,x_t}(h)\big)$. Since $\lambda_1^{q_n,x_t}(h) \to \lambda_1^{f,x_t}(h)$ for all $(x_0, \ldots, x_t) \in \mathsf{X}^{t+1}$ by Assumption 3.1-(b) and $\lambda_{(t)}^{q_n,x} \to \lambda_{(t)}^{f,x}$ setwise, we have $\lambda_{(t+1)}^{q_n,x}(h) \to \lambda_{(t+1)}^{f,x}(h)$ again by [122, Theorem 2.4] which completes the proof.

Proof of Lemma 3.17
Let $Q(\cdot|x,a)$ be the nonnegative measure on X defined by $Q(D|x,a) = \int_D w(y)p(dy|x,a)$ for $D \in \mathscr{B}(\mathsf{X})$ and let Ξ be the family of measures

$$\Xi := \biggl\{ Q(\cdot|x,a) : (x,a) \in K \times \mathsf{A} \biggr\}.$$

Note that Ξ is uniformly bounded since

$$\sup_{(x,a)\in K\times\mathsf{A}} \int_{\mathsf{X}} w(y)p(dy|x,a) \leq \alpha \sup_{x\in K} w(x) < \infty.$$

If the mapping $Q : K \times \mathsf{A} \ni (x,a) \mapsto Q(\cdot|x,a) \in \mathscr{M}_+(\mathsf{X})$ is continuous with respect to the weak topology on $\mathscr{M}_+(\mathsf{X})$, then Ξ (being a continuous image of the compact set $K \times \mathsf{A}$) is compact with respect to the weak topology. Then, by Prohorov's theorem [23, Theorem 8.6.2], Ξ is tight, completing the proof. Hence, we only need to prove the continuity of the mapping Q.

By Lemma 3.13, for any $u \in C_w(\mathsf{X})$, $\int_{\mathsf{X}} u(y)p(dy|x,a)$ is continuous in (x,a). Let $(x_k,a_k) \to (x,a)$ in $K \times \mathsf{A}$. Note that for any $g \in C_b(\mathsf{X})$, $gw \in C_w(\mathsf{X})$. Therefore, we have

$$\begin{aligned}\lim_{k\to\infty} \int_{\mathsf{X}} g(y)Q(dy|x_k,a_k) &= \lim_{k\to\infty} \int_{\mathsf{X}} g(y)w(y)p(dy|x_k,a_k) \\ &= \int_{\mathsf{X}} g(y)w(y)p(dy|x,a) = \int_{\mathsf{X}} g(y)Q(dy|x,a)\end{aligned}$$

proving that $Q(\cdot|x_k,a_k) \to Q(\cdot|x,a)$ weakly.

Proof of Lemma 3.18
Fix any compact subset K of X. For any $\varepsilon > 0$, let $K_\varepsilon \subset \mathsf{X}$ be the compact set in Lemma 3.17 for K. We have

$$\begin{aligned}&\sup_{x\in K} \biggl| \int_{\mathsf{X}} u_n(y)p(dy|x,f_n(x)) - \int_{\mathsf{X}} u(y)p(dy|x,f(x)) \biggr| \\ &\quad\leq \sup_{x\in K} \biggl| \int_{\mathsf{X}} u_n(y)p(dy|x,f_n(x)) - \int_{\mathsf{X}} u(y)p(dy|x,f_n(x)) \biggr| \\ &\qquad + \sup_{x\in K} \biggl| \int_{\mathsf{X}} u(y)p(dy|x,f_n(x)) - \int_{\mathsf{X}} u(y)p(dy|x,f(x)) \biggr| \\ &\quad\leq \sup_{x\in K} \biggl| \int_{K_\varepsilon} u_n(y)p(dy|x,f_n(x)) - \int_{K_\varepsilon} u(y)p(dy|x,f_n(x)) \biggr|\end{aligned}$$

$$+ \sup_{x \in K} \left| \int_{K_\varepsilon^c} u_n(y) p(dy|x, f_n(x)) - \int_{K_\varepsilon^c} u(y) p(dy|x, f_n(x)) \right|$$

$$+ \sup_{x \in K} \left| \int_{\mathsf{X}} u(y) p(dy|x, f_n(x)) - \int_{\mathsf{X}} u(y) p(dy|x, f(x)) \right|$$

$$\leq \sup_{y \in K_\varepsilon} |u_n(y) - u(y)| + (L + \|u\|_w)\varepsilon \text{ (by Lemma 3.17)}$$

$$+ \sup_{x \in K} \left| \int_{\mathsf{X}} u(y) p(dy|x, f_n(x)) - \int_{\mathsf{X}} u(y) p(dy|x, f(x)) \right|.$$

Let us define $l(x, a) := \int_{\mathsf{X}} u(y) p(dy|x, a)$. Since $u(y) \in C_w(\mathsf{X})$, by Lemma 3.13 we have that l is continuous, and therefore, it is uniformly continuous on $K \times \mathsf{A}$. Note that in the last expression, as $n \to \infty$ the first term goes to zero since $u_n \to u$ uniformly on K_ε and the last term goes to zero since l is uniformly continuous on $K \times \mathsf{A}$ and $f_n \to f$ uniformly. Then the result follows since ε is arbitrary.

Proof of Proposition 3.28

We will prove this result by induction. Let x be an arbitrary initial point and fix n. For $t = 1$ the claim holds by the following argument:

$$\begin{aligned}
\|\lambda_1^{f,x} - \lambda_1^{q_n,x}\|_{TV} &\leq K_2 d_{\mathsf{A}}(f(x), q_n(x)) \text{ (by Assumption 3.5-(k))} \\
&\leq (1/n)^{1/d} K_2 \alpha \text{ (by Lemma 3.27).}
\end{aligned}$$

Observe that the bound $\alpha K_2 (2t - 1)(1/n)^{1/d}$ is independent of the choice of initial point x for $t = 1$. Assume the claim is true for $t \geq 1$. Note that for any $g \in \mathbb{F}$ and $h \in B(\mathsf{X})$, we can write

$$\begin{aligned}
\lambda_{t+1}^{g,x}(h) &= \int_{\mathsf{X}} h(x_{t+1}) P_x^g(dx_{t+1}) \\
&= \int_{\mathsf{X}} \int_{\mathsf{X}} h(x_t) P_{x_1}^g(dx_t) P_x^g(dx_1) \text{ (by the Markov property)} \\
&= \lambda_1^{g,x}(\lambda_t^{g,x_1}(h)).
\end{aligned}$$

Then we have

$$\begin{aligned}
\|\lambda_{t+1}^{f,x} - \lambda_{t+1}^{q_n,x}\|_{TV} &= 2 \sup_{B \in \mathscr{B}(\mathsf{X})} \left| \lambda_1^{f,x}(\lambda_t^{f,x_1}(B)) - \lambda_1^{q_n,x}(\lambda_t^{q_n,x_1}(B)) \right| \\
&= 2 \sup_{B \in \mathscr{B}(\mathsf{X})} \left| \lambda_1^{f,x}(\lambda_t^{f,x_1}(B)) - \lambda_1^{f,x}(\lambda_t^{q_n,x_1}(B)) + \lambda_1^{f,x}(\lambda_t^{q_n,x_1}(B)) - \lambda_1^{q_n,x}(\lambda_t^{q_n,x_1}(B)) \right| \\
&\leq \lambda_1^{f,x}(\|\lambda_t^{f,x_1} - \lambda_t^{q_n,x_1}\|_{TV}) + 2\|\lambda_1^{f,x} - \lambda_1^{q_n,x}\|_{TV} \qquad (3.29)
\end{aligned}$$

$$\leq (1/n)^{1/d}(2t-1)K_2\alpha + 2(1/n)^{1/d}K_2\alpha \tag{3.30}$$
$$= \alpha K_2(2(t+1)-1)(1/n)^{1/d}.$$

Here (3.29) follows since $|\mu(h) - \eta(h)| \leq \|\mu - \eta\|_{TV}\|h\|$ and (3.30) follows since the bound $\lambda K_2(2t-1)(1/n)^{1/d}$ is independent of the initial distribution.

3.7 Concluding Remarks

In this chapter, we considered the finite-action approximation of stationary policies for a discrete-time Markov decision process with discounted and average costs. Under mild technical assumptions, it was shown that if one uses a sufficiently large number of points to discretize the action space, then the resulting finite-action MDP can approximate the original model with arbitrary precision. Under the Lipschitz continuity of the transition probability and the one-stage cost function, explicit bounds were obtained on the performance loss due to quantization. The results of this chapter are based on the papers [112–114, 118].

The investigation of the finite-action approximation problem is also useful in learning algorithms for MDPs. If one can ensure that learning a control model with only finitely many control actions is sufficient for approximate optimality, then it is easier to develop efficient learning algorithms which allow for the approximate computation of finitely many transition probabilities. In particular, results developed in the learning and information theory literature for conditional kernel estimations [61] (with control-free models) can be applied to transition probability estimation for MDPs. It would also be interesting to establish similar results for approximations where the set of admissible quantizers has a certain structure, such as the set of quantizers having convex codecells [62], which may give rise to practical design methods.

Chapter 4
Finite-State Approximation of Markov Decision Processes

4.1 Introduction

In this chapter we study the finite-state approximation problem for computing near optimal policies for discrete-time MDPs with Borel state and action spaces, under discounted and average costs criteria. Even though existence and structural properties of optimal policies of MDPs have been studied extensively in the literature, computing such policies is generally a challenging problem for systems with uncountable state spaces. This situation also arises in the fully observed reduction of a partially observed Markov decision process even when the original system has finite state and action spaces. Here we show that one way to compute approximately optimal solutions for such MDPs is to construct a reduced model with a new transition probability and one-stage cost function by quantizing the state space, i.e., by discretizing it on a finite grid. It is reasonable to expect that when the one-stage cost function and the transition probability of the original model have certain continuity properties, the cost of the optimal policy for the approximating finite model converges to the optimal cost of the original model as the discretization becomes finer. Moreover, under additional continuity conditions on the transition probability and the one stage cost function we also obtain bounds on the accuracy of the approximation in terms of the number of points used to discretize the state space, thereby providing a tradeoff between the computation cost and the performance loss in the system. In particular, we study the following two problems.

(Q1) Under what conditions on the components of the MDP do the true cost functions of the policies obtained from finite models converge to the optimal value function as the number of grid points goes to infinity? Here, we are only concerned with the convergence of the approximation and not with bounds on the rate of approximation.

N. Saldi et al., *Finite Approximations in Discrete-Time Stochastic Control*,
Systems & Control: Foundations & Applications,
https://doi.org/10.1007/978-3-319-79033-6_4

(Q2) Under what conditions can we obtain bounds on the performance loss due to discretization in terms of the number of grid points if we strengthen the conditions sufficient in **(Q1)**?

Our approach to problem **(Q1)** can be summarized as follows: (i) first, we obtain approximation results for the compact-state case, (ii) we find conditions under which a compact representation leads to near optimality for non-compact state MDPs, and (iii) we prove the convergence of the finite-state models to non-compact models. As a by-product of this analysis, we obtain *compact-state-space approximations* for an MDP with non-compact Borel state space. In particular, our results directly lead to finite models if the state space is countable; similar problems in the countable context have been studied in the literature for the discounted cost [104, Section 6.10.2].

Combined with Chapter 3 where we investigated the asymptotic optimality of the quantization of action sets, the results in this chapter lead to a constructive algorithm for obtaining approximately optimal solutions. First the action space is quantized with small error, and then the state space is quantized with small error, which results in a finite model that well approximates the original MDP. When the state space is compact, we can also obtain rates of convergence for both approximations, and using information theoretic tools we can establish that the obtained rates of convergence are order-optimal for a given class of MDPs.

4.1.1 Auxiliary Results

To avoid measurability problems associated with the operators that will be defined for the approximation problem in the discounted cost case, it is necessary to enlarge the set of functions on which these operators act. To this end, in this section we review the notion of analytic sets and lower semi-analytic functions, and state the main results that will be used in the sequel to tackle these measurability problems. For a detailed treatment of analytic sets and lower semi-analytic functions, we refer the reader to [21, 123], [83, Chapter 39], and [16, Chapter 7].

Let $\mathbb{N}^\infty$ be the set of sequences of natural numbers endowed with the product topology. With this topology, $\mathbb{N}^\infty$ is a complete and separable metric space. A subset A of a Borel space E is said to be *analytic* if it is a continuous image of $\mathbb{N}^\infty$. Note that Borel sets are always analytic.

A function $g : \mathsf{E} \to \mathbb{R}$ is said to be *universally measurable* if for any $\mu \in \mathscr{P}(\mathsf{E})$, there is a Borel measurable function $g_\mu : \mathsf{E} \to \mathbb{R}$ such that $g = g_\mu$ μ-almost everywhere. It is said to be *lower semi-analytic* if the set $\{e : g(e) < c\}$ is analytic for any $c \in \mathbb{R}$. Any Borel measurable function is lower semi-analytic and any lower semi-analytic function is universally measurable. The latter property implies that the integral of any lower semi-analytic function with respect to any probability measure is well defined. We let $B^l(\mathsf{E})$ and $B^l_w(\mathsf{E})$ denote the set of all bounded lower semi-analytic functions and lower semi-analytic functions with finite

w-norm, respectively. Since any pointwise limit of a sequence of lower semi-analytic functions is lower semi-analytic (see [83, Theorem 1, p. 512]), $(B^l(\mathsf{E}), \|\cdot\|)$ and $(B^l_w(\mathsf{E}), \|\cdot\|_w)$ are Banach spaces.

We now state the results that will be used in the sequel.

Proposition 4.1 ([16, Proposition 7.47]) *Suppose* E_1 *and* E_2 *are Borel spaces. Let* $g : \mathsf{E}_1 \times \mathsf{E}_2 \to \mathbb{R}$ *be lower semi-analytic. Then,* $g^*(e_1) := \inf_{e_2 \in \mathsf{E}_2} g(e_1, e_2)$ *is also lower semi-analytic.*

Proposition 4.2 ([16, Proposition 7.48]) *Suppose* E_1 *and* E_2 *are as in Proposition 4.1. Let* $g : \mathsf{E}_1 \times \mathsf{E}_2 \to \mathbb{R}$ *be lower semi-analytic and let* $q(de_2|e_1)$ *be a stochastic kernel on* E_2 *given* E_1*. Then, the function*

$$h(e_1) := \int_{\mathsf{E}_2} g(e_1, e_2) q(de_2|e_1).$$

is lower semi-analytic.

4.2 Finite-State Approximation of Compact State MDPs

In this section we consider **(Q1)** for the MDPs with compact state space. To distinguish compact-state MDPs from non-compact ones, the state space of the compact-state MDPs will be denoted by Z instead of X. We impose the assumptions below on the components of the Markov decision process; additional new assumptions will be made for the average cost problem in Section 4.2.2.

Assumption 4.1

(a) The one-stage cost function c *is continuous and bounded.*
(b) The stochastic kernel $p(\cdot|z, a)$ *is weakly continuous in* (z, a)*.*
(c) Z *and* A *are compact.*

Analogous with the construction of finite-action models in Section 3.2.1, we first describe the procedure used to obtain finite-state models. Let d_{Z} denote the metric on Z. Since the state space Z is compact and thus totally bounded, there exists a sequence $\big(\{z_{n,i}\}_{i=1}^{k_n}\big)_{n \geq 1}$ of finite grids in Z such that for all n,

$$\min_{i \in \{1,\ldots,k_n\}} d_{\mathsf{Z}}(z, z_{n,i}) < 1/n \text{ for all } z \in \mathsf{Z}.$$

Let $\mathsf{Z}_n := \{z_{n,1}, \ldots, z_{n,k_n}\}$ and define function $Q_n : \mathsf{Z} \to \mathsf{Z}_n$ by

$$Q_n(z) := \arg\min_{y \in \mathsf{Z}_n} d_{\mathsf{Z}}(z, y),$$

where ties are broken so that Q_n is measurable. In the literature, Q_n is often called a nearest neighbor quantizer with respect to "distortion measure" d_{Z} [58]. For each n, Q_n induces a partition $\{\mathcal{S}_{n,i}\}_{i=1}^{k_n}$ of the state space Z given by

$$\mathcal{S}_{n,i} = \{z \in \mathsf{Z} : Q_n(z) = z_{n,i}\},$$

with diameter $\mathrm{diam}(\mathcal{S}_{n,i}) := \sup_{z,y \in \mathcal{S}_{n,i}} d_{\mathsf{Z}}(z, y) < 2/n$. Let $\{\nu_n\}$ be a sequence of probability measures on Z satisfying

$$\nu_n(\mathcal{S}_{n,i}) > 0 \text{ for all } i, n. \tag{4.1}$$

We let $\nu_{n,i}$ be the restriction of ν_n to $\mathcal{S}_{n,i}$ defined by

$$\nu_{n,i}(\,\cdot\,) := \frac{\nu_n(\,\cdot\,)}{\nu_n(\mathcal{S}_{n,i})}.$$

The measures $\nu_{n,i}$ will be used to define a sequence of finite-state MDPs, denoted as MDP_n, which approximate the original model. To this end, for each n define the one-stage cost function $c_n : \mathsf{Z}_n \times \mathsf{A} \to [0, \infty)$ and the transition probability p_n on Z_n given $\mathsf{Z}_n \times \mathsf{A}$ by

$$c_n(z_{n,i}, a) := \int_{\mathcal{S}_{n,i}} c(z, a)\nu_{n,i}(dz),$$

$$p_n(\,\cdot\,|z_{n,i}, a) := \int_{\mathcal{S}_{n,i}} Q_n * p(\,\cdot\,|z, a)\nu_{n,i}(dz),$$

where $Q_n * p(\,\cdot\,|z, a) \in \mathcal{P}(\mathsf{Z}_n)$ is the pushforward of the measure $p(\,\cdot\,|z, a)$ with respect to Q_n; that is,

$$Q_n * p(y|z, a) = p\big(\{z \in \mathsf{Z} : Q_n(z) = y\}|z, a\big),$$

for all $y \in \mathsf{Z}_n$. For each n, we define MDP_n as a Markov decision process with the following components: Z_n is the state space, A is the action space, p_n is the transition probability, and c_n is the one-stage cost function. History spaces, policies, and cost functions are defined in a similar way as in the original model.

4.2.1 *Discounted Cost*

In this section we suppose that Assumption 4.1 holds. Recall the Bellman optimality operator T defined in (3.9). It can be proved that under Assumption 4.1-(a)(b), T is a contraction operator with modulus β mapping $C_b(\mathsf{Z})$ into itself (see [66, Theorem 2.8]); that is, $Tu \in C_b(\mathsf{Z})$ for all $u \in C_b(\mathsf{Z})$ and

$$\|Tu\| \leq \beta \|u\| \text{ for all } u \in C_b(\mathsf{Z}).$$

Since Assumption 3.3 holds with $w \equiv 1$ in this case, Theorem 3.15 in Section 3.4.1 implies that the value function J^* is the unique fixed point in $C_b(\mathsf{Z})$ of the contraction operator T, i.e.,

$$J^* = TJ^*.$$

Furthermore, a deterministic stationary policy f^* is optimal if and only if it satisfies the optimality equation, i.e.,

$$J^*(z) = c(z, f^*(z)) + \beta \int_{\mathsf{Z}} J^*(y) p(dy|z, f^*(z)). \tag{4.2}$$

Finally, there exists a deterministic stationary policy f^* which is optimal, so it satisfies (4.2).

For any $n \geq 1$ define the operator T_n (Bellman optimality operator for MDP_n) by

$$T_n u(z_{n,i}) := \min_{a \in \mathsf{A}} \left[c_n(z_{n,i}, a) + \beta \sum_{j=1}^{k_n} u(z_{n,j}) p_n(z_{n,j}|z_{n,i}, a) \right],$$

or equivalently,

$$T_n u(z_{n,i}) = \min_{a \in \mathsf{A}} \int_{\mathscr{S}_{n,i}} \left[c(z, a) + \beta \int_{\mathsf{Z}} \hat{u}(y) p(dy|z, a) \right] \nu_{n,i}(dz),$$

where $u : \mathsf{Z}_n \to \mathbb{R}$ and $\hat{u}$ is the piecewise constant extension of u to Z given by $\hat{u}(z) = u \circ Q_n(z)$. For each n, [66, Theorem 2.8] implies the following: (i) T_n is a contraction operator with modulus β mapping $B(\mathsf{Z}_n)$ $\big(= C_b(\mathsf{Z}_n)\big)$ into itself, (ii) the fixed point of T_n is the value function J_n^* of MDP_n, and (iii) there exists an optimal stationary policy f_n^* for MDP_n, which therefore satisfies the optimality equation. Hence, we have

$$J_n^* = T_n J_n^* = T_n J_n(f_n^*, \cdot) = J_n(f_n^*, \cdot),$$

where J_n denotes the discounted cost for MDP_n. Let us extend the optimal policy f_n^* for MDP_n to Z by letting $\hat{f}_n(z) = f_n^* \circ Q_n(z) \in \mathbb{F}$.

The following theorem is the main result of this section. It states that the cost function of the policy $\hat{f}_n$ converges to the original value function J^* as $n \to \infty$.

Theorem 4.3 *The discounted cost of the policy $\hat{f}_n$, obtained by extending the optimal policy f_n^* of MDP$_n$ to Z, converges to the optimal value function J^* of the original MDP*

$$\lim_{n\to\infty} \|J(\hat{f}_n, \cdot) - J^*\| = 0.$$

Hence, to find a near optimal policy for the original MDP, it is sufficient to compute the optimal policy of MDP$_n$ for sufficiently large n, and then extend this policy to the original state space.

To prove Theorem 4.3 we need a series of technical results. We first define an operator $\hat{T}_n$ on $B^l(\mathsf{Z})$ by extending T_n to $B^l(\mathsf{Z})$:

$$\hat{T}_n u(z) := \inf_{a\in\mathsf{A}} \int_{\mathscr{S}_{n,i_n(z)}} \biggl[c(x,a) + \beta \int_{\mathsf{Z}} u(y) p(dy|x,a)\biggr] \nu_{n,i_n(z)}(dx), \tag{4.3}$$

where $i_n : \mathsf{Z} \to \{1,\ldots,k_n\}$ maps z to i where $Q_n(z) = z_{n,i}$. To see that this operator is well defined, let the stochastic kernel $r_n(dx|z)$ on Z given Z be defined as

$$r_n(dx|z) := \sum_{i=1}^{k_n} \nu_{n,i}(dx) 1_{\mathscr{S}_{n,i}}(z).$$

Then, we can write the right-hand side of (4.3) as

$$\inf_{a\in\mathsf{A}} \int_{\mathsf{Z}} \biggl[c(x,a) + \beta \int_{\mathsf{Z}} u(y) p(dy|x,a)\biggr] r_n(dx|z).$$

Therefore, by Propositions 4.1 and 4.2, we can conclude that $\hat{T}_n$ maps $B^l(\mathsf{Z})$ into $B^l(\mathsf{Z})$. Furthermore, it is a contraction operator with modulus β which can be shown using [66, Proposition A.2]. Hence, it has a unique fixed point $\hat{J}_n^*$ that belongs to $B^l(\mathsf{Z})$, and this fixed point must be constant over the sets $\mathscr{S}_{n,i}$ because of the averaging operation on each $\mathscr{S}_{n,i}$. Furthermore, since $\hat{T}_n(u \circ Q_n) = (T_n u) \circ Q_n$ for all $u \in B(\mathsf{Z}_n)$, we have

$$\hat{T}_n(J_n^* \circ Q_n) = (T_n J_n^*) \circ Q_n = J_n^* \circ Q_n.$$

Hence, the fixed point of $\hat{T}_n$ is the piecewise constant extension of the fixed point of T_n, i.e.,

$$\hat{J}_n^* = J_n^* \circ Q_n \in B(\mathsf{Z}).$$

Remark 4.4 In the rest of this chapter, when we take the integral of any function with respect to $\nu_{n,i_n(z)}$, it is tacitly assumed that the integral is taken over $\mathscr{S}_{n,i_n(z)}$. Hence, we can drop $\mathscr{S}_{n,i_n(z)}$ in the integral to simplify the notation.

We now define another operator F_n on $B^l(\mathsf{Z})$ by simply interchanging the order of the minimum and the integral in (4.3), i.e.,

$$F_n u(z) := \int \min_{a \in \mathsf{A}} \left[c(x, a) + \beta \int_{\mathsf{Z}} u(y) p(dy|x, a) \right] \nu_{n,i_n(z)}(dx)$$
$$= \Gamma_n T u(z),$$

where

$$\Gamma_n u(z) := \int u(x) \nu_{n,i_n(z)}(dx).$$

We note that F_n is the extension (to infinite state spaces) of the operator defined in [110, p. 236] for an approximate value iteration algorithm. However, unlike in [110], F_n will serve here as an intermediate step between T and $\hat{T}_n$ (or T_n) to solve **(Q1)** for the discounted cost. To this end, we first note that F_n is a contraction operator on $B^l(\mathsf{Z})$ with modulus β. Indeed it is clear that F_n maps $B^l(\mathsf{Z})$ into itself by Propositions 4.1 and 4.2. Furthermore, for any $u, v \in B^l(\mathsf{Z})$, we clearly have $\|\Gamma_n u - \Gamma_n v\| \leq \|u - v\|$. Hence, since T is a contraction operator on $B^l(\mathsf{Z})$ with modulus β, F_n is also a contraction operator on $B^l(\mathsf{Z})$ with modulus β.

Remark 4.5 Since we only assume that the stochastic kernel p is weakly continuous, it is not true that $\hat{T}_n$ and F_n map $B(\mathsf{Z})$ into itself (see [68, Proposition D.5, p. 182]). This is the point where we need to enlarge the set of functions on which these operators act.

The following theorem states that the fixed point, say u_n^*, of F_n converges to the fixed point J^* of T (i.e., the value function) as $n \to \infty$. Note that although T is originally defined on $C_b(\mathsf{Z})$, it can be proved that when acting on $B^l(\mathsf{Z})$, it maps $B^l(\mathsf{Z})$ into itself.

Theorem 4.6 *If u_n^* is the unique fixed point of F_n, then* $\lim_{n \to \infty} \|u_n^* - J^*\| = 0$.

The proof of Theorem 4.6 requires two lemmas.

Lemma 4.7 *For any $u \in B^l(\mathsf{Z})$, we have*

$$\|u - \Gamma_n u\| \leq 2 \inf_{r \in \mathsf{Z}^{k_n}} \|u - \Phi_r\|,$$

where $\Phi_r(z) = \Sigma_{i=1}^{k_n} r_i 1_{\mathscr{S}_{n,i}}(z)$, $r = (r_1, \cdots, r_{k_n})$.

Proof Fix any $r \in \mathsf{Z}^{k_n}$. Then, using the identity $\Gamma_n \Phi_r = \Phi_r$, we obtain

$$\begin{aligned}\|u - \Gamma_n u\| &\leq \|u - \Phi_r\| + \|\Phi_r - \Gamma_n u\| \\ &= \|u - \Phi_r\| + \|\Gamma_n \Phi_r - \Gamma_n u\| \\ &\leq \|u - \Phi_r\| + \|\Phi_r - u\|.\end{aligned}$$

Since r is arbitrary, this completes the proof. □

Notice that by the definition of the operator Γ_n, the fixed point u_n^* of F_n is constant over the sets $\mathcal{S}_{n,i}$. We use this property to prove the next lemma.

Lemma 4.8 *We have*

$$\|u_n^* - J^*\| \leq \frac{2}{1-\beta} \inf_{r \in \mathsf{Z}^{k_n}} \|J^* - \Phi_r\|.$$

Proof Note that $\Gamma_n u_n^* = u_n^*$ since u_n^* is constant over each $\mathcal{S}_{n,i}$. Then, we have

$$\begin{aligned}\|u_n^* - J^*\| &\leq \|u_n^* - \Gamma_n J^*\| + \|\Gamma_n J^* - J^*\| \\ &= \|F_n u_n^* - \Gamma_n T J^*\| + \|\Gamma_n J^* - J^*\| \\ &= \|\Gamma_n T u_n^* - \Gamma_n T J^*\| + \|\Gamma_n J^* - J^*\| \text{ (by the definition of } F_n) \\ &\leq \|T u_n^* - T J^*\| + \|\Gamma_n J^* - J^*\| \text{ (since} \|\Gamma_n u - \Gamma_n v\| \leq \|u - v\|) \\ &\leq \beta \|u_n^* - J^*\| + \|\Gamma_n J^* - J^*\|.\end{aligned}$$

Hence, we obtain $\|u_n^* - J^*\| \leq \frac{1}{1-\beta} \|\Gamma_n J^* - J^*\|$. The result now follows from Lemma 4.7. □

Proof of Theorem 4.6. Recall that since Z is compact, the function J^* is uniformly continuous and $\mathrm{diam}(\mathcal{S}_{n,i}) < 2/n$ for all $i = 1, \ldots, k_n$. Hence, $\inf_{r \in \mathsf{Z}^{k_n}} \|J^* - \Phi_r\| \to 0$ as $n \to \infty$, which completes the proof in view of Lemma 4.8. □

The next step is to show that the fixed point $\hat{J}_n^*$ of $\hat{T}_n$ converges to the fixed point J^* of T. To this end, we first prove the following result.

Lemma 4.9 *For any $u \in C_b(\mathsf{Z})$, $\|\hat{T}_n u - F_n u\| \to 0$ as $n \to \infty$.*

Proof Note that since $\int_{\mathsf{Z}} u(x) p(dx|y,a)$ is continuous as a function of (y,a) by Assumption 4.1-(b), it is sufficient to prove that for any $l \in C_b(\mathsf{Z} \times \mathsf{A})$

$$\begin{aligned}&\left\| \min_a \int l(y,a) \nu_{n,i_n(z)}(dy) - \int \min_a l(y,a) \nu_{n,i_n(z)}(dy) \right\| \\ &\qquad = \sup_{z \in \mathsf{Z}} \left| \min_a \int l(y,a) \nu_{n,i_n(z)}(dy) - \int \min_a l(y,a) \nu_{n,i_n(z)}(dy) \right| \to 0\end{aligned}$$

as $n \to \infty$. Fix any $\varepsilon > 0$. Define $\{z_i\}_{i=1}^{\infty} := \bigcup_n \mathsf{Z}_n$ and let $\{a_i\}_{i=1}^{\infty}$ be a sequence in A such that $\min_{a \in \mathsf{A}} l(z_i, a) = l(z_i, a_i)$; such a_i exists for each z_i because $l(z_i, \cdot)$ is continuous and A is compact. Define $g(y) := \min_{a \in \mathsf{A}} l(y,a)$, which can be proved to be continuous, and therefore uniformly continuous since Z is compact. Thus by the uniform continuity of l, there exists $\delta > 0$ such that $d_{\mathsf{Z}\times\mathsf{A}}\big((y,a),(y',a')\big) < \delta$ implies $|g(y) - g(y')| < \varepsilon/2$ and $|l(y,a) - l(y',a')| < \varepsilon/2$. Choose n_0 such that $2/n_0 < \delta$. Then for all $n \geq n_0$, $\max_{i \in \{1,\ldots,k_n\}} \operatorname{diam}(\mathscr{S}_{n,i}) < 2/n < \delta$. Hence, for all $y \in \mathscr{S}_{n,i}$ we have $|l(y,a_i) - \min_{a \in \mathsf{A}} l(y,a)| \leq |l(y,a_i) - l(z_i,a_i)| + |\min_{a \in \mathsf{A}} l(z_i,a) - \min_{a \in \mathsf{A}} l(y,a)| = |l(y,a_i) - l(z_i,a_i)| + |g(z_i) - g(y)| < \varepsilon$. This implies

$$\begin{aligned}
&\bigg\| \min_a \int l(y,a)\nu_{n,i_n(z)}(dy) - \int \min_a l(y,a)\nu_{n,i_n(z)}(dy) \bigg\| \\
&\quad \leq \bigg\| \int l(y,a_i)\nu_{n,i_n(z)}(dy) - \int \min_a l(y,a)\nu_{n,i_n(z)}(dy) \bigg\| \\
&\quad \leq \sup_{z \in \mathsf{Z}} \int \sup_{y \in \mathscr{S}_{n,i_n(z)}} \big|l(y,a_i) - \min_a l(y,a)\big| \nu_{n,i_n(z)}(dy) < \varepsilon.
\end{aligned}$$

This completes the proof. □

Theorem 4.10 *The fixed point $\hat{J}_n^*$ of $\hat{T}_n$ converges to the fixed point J^* of T.*

Proof We have

$$\begin{aligned}
\|\hat{J}_n^* - J^*\| &\leq \|\hat{T}_n \hat{J}_n^* - \hat{T}_n J^*\| + \|\hat{T}_n J^* - F_n J^*\| + \|F_n J^* - F_n u_n^*\| + \|F_n u_n^* - J^*\| \\
&\leq \beta \|\hat{J}_n^* - J^*\| + \|\hat{T}_n J^* - F_n J^*\| + \beta \|J^* - u_n^*\| + \|u_n^* - J^*\|.
\end{aligned}$$

Hence

$$\|\hat{J}_n^* - J^*\| \leq \frac{\|\hat{T}_n J^* - F_n J^*\| + (1+\beta)\|J^* - u_n^*\|}{1-\beta}.$$

The theorem now follows from Theorem 4.6 and Lemma 4.9. □

Recall the optimal stationary policy f_n^* for MDP_n and its extension $\hat{f}_n(z) = f_n^* \circ Q_n(z)$ to Z. Since $\hat{J}_n^* = J_n^* \circ Q_n$, it is straightforward to prove that $\hat{f}_n$ is the optimal selector of $\hat{T}_n \hat{J}_n^*$; that is,

$$\hat{T}_n \hat{J}_n^* = \hat{J}_n^* = \hat{T}_{\hat{f}_n} \hat{J}_n^*,$$

where $\hat{T}_{\hat{f}_n}$ is defined as

$$\hat{T}_{\hat{f}_n} u(z) := \int \bigg[c(x, \hat{f}_n(x)) + \beta \int_{\mathsf{Z}} u(y) p(dy|x, \hat{f}_n(x)) \bigg] \nu_{n,i_n(z)}(dx).$$

Define analogously

$$T_{\hat{f}_n} u(z) := c(z, \hat{f}_n(z)) + \beta \int_{\mathsf{Z}} u(y) p(dy|z, \hat{f}_n(z)).$$

It can be proved that both $\hat{T}_{\hat{f}_n}$ and $T_{\hat{f}_n}$ are contraction operators on $B^l(\mathsf{Z})$ with modulus β, and it is known that the fixed point of $T_{\hat{f}_n}$ is the true cost function of the stationary policy $\hat{f}_n$ (i.e., $J(\hat{f}_n, z)$).

Lemma 4.11 $\|\hat{T}_{\hat{f}_n} u - T_{\hat{f}_n} u\| \to 0$ *as* $n \to \infty$, *for any* $u \in C_b(\mathsf{Z})$.

Proof The statement follows from the uniform continuity of the function $c(z, a) + \beta \int_{\mathsf{Z}} u(y) p(dy|z, a)$ and the fact that $\hat{f}_n$ is constant over the sets $\mathscr{S}_{n,i}$. □

Now, we prove the main result of this section.

Proof of Theorem 4.3. We have

$$\begin{aligned}
&\|J(\hat{f}_n, \cdot) - J^*\| \\
&\leq \|T_{\hat{f}_n} J(\hat{f}_n, \cdot) - T_{\hat{f}_n} J^*\| + \|T_{\hat{f}_n} J^* - \hat{T}_{\hat{f}_n} J^*\| + \|\hat{T}_{\hat{f}_n} J^* - \hat{T}_{\hat{f}_n} \hat{J}_n^*\| + \|\hat{J}_n^* - J^*\| \\
&\leq \beta \|J(\hat{f}_n, \cdot) - J^*\| + \|T_{\hat{f}_n} J^* - \hat{T}_{\hat{f}_n} J^*\| + \beta \|J^* - \hat{J}_n^*\| + \|\hat{J}_n^* - J^*\|.
\end{aligned}$$

Hence, we obtain

$$\|J(\hat{f}_n, \cdot) - J^*\| \leq \frac{\|T_{\hat{f}_n} J^* - \hat{T}_{\hat{f}_n} J^*\| + (1 + \beta)\|\hat{J}_n^* - J^*\|}{1 - \beta}.$$

The result follows from Lemma 4.11 and Theorem 4.10. □

4.2.2 Average Cost

Here we need to impose some new conditions on the components of the original MDP in addition to Assumption 4.1.

Assumption 4.2 *Suppose Assumption 4.1 holds with item (b) replaced by the condition that the stochastic kernel* $p(\cdot|z, a)$ *is continuous in* (z, a) *with respect to the total variation distance. In addition, suppose there exist* $\lambda \in \mathscr{P}(\mathsf{Z})$, $\alpha \in (0, 1)$, *and* $\phi \in B(\mathsf{Z} \times \mathsf{A})$ *such that*

(d) $p(D|z, a) \geq \lambda(D)\phi(z, a)$ *for all* $D \in \mathscr{B}(\mathsf{Z})$,
(e) $1 - \alpha \leq \phi(z, a)$.

Throughout this section, it is assumed that Assumption 4.2 holds. Recall that any deterministic stationary policy f defines a stochastic kernel $p(\cdot|z, f(z))$ on Z given

Z which is the transition probability of the Markov chain $\{Z_t\}_{t=0}^{\infty}$ (state process) induced by f. In this chapter, instead of the notation introduced in Section 3.3.2, we write $p^t(\,\cdot\,|z, f(z))$ to denote the t-step transition probability of this Markov chain given the initial point z. This is done in order to distinguish $p^t(\,\cdot\,|z, f(z))$ from $p^t(\,\cdot\,|z, f(y))$ for $z \neq y$.

To study average cost optimal control problems, it is usually assumed that there exists an invariant distribution under any stationary control policy, so that the average cost of any stationary policy can be written as an integral of the one-stage cost function with respect to this invariant distribution. With this representation, one can then deduce the optimality of stationary policies using the linear programming or the convex analytic methods (see [24, 68]). However, to solve the approximation problem for the average cost, we need, in addition to the existence of an invariant distribution, the convergence of the t-step transition probabilities to the invariant distribution, at some explicitly given rate, for both the original and the reduced problems. Therefore, it is crucial to impose appropriate conditions on the original model such that, on the one hand, they guarantee the convergence of t-step transition probabilities to the invariant distribution for all stationary policies for the original system and, on the other hand, one is able to show that similar conditions are satisfied by the reduced problems. Conditions (d) and (e) in Assumption 4.2 are examples of such conditions which have also been used in the literature extensively. Indeed, if we define the weight function $w \equiv 1$, then condition (e) corresponds to the so-called "drift inequality": for all $(z, a) \in \mathsf{Z} \times \mathsf{A}$

$$\int_{\mathsf{Z}} w(y) p(dy|z, a) \leq \alpha w(z) + \lambda(w) \phi(z, a),$$

and condition (d) corresponds to the so-called "minorization" condition, both of which have been used in literature for studying geometric ergodicity of Markov chains (see Section 3.3.2, and references therein).

We refer the reader to [109, Theorem 9], [96, Theorem 15.0.1], and [69] for alternative conditions that imply geometric ergodicity of Markov chains induced by stationary policies and which can replace conditions (d) and (e) in Assumption 4.2. For sake of completeness, it is useful to note that the geometric ergodicity conditions can be relaxed for some of the results in this section. Further related results, on subgeometric rates of convergence, are available [64, Sec. 4], [32, 38, 127, Sec. 5].

The following theorem is a consequence of [129, Theorem 3.3], [55, Lemma 3.4], and [78, Theorem 3], which also holds when the continuity of $p(\,\cdot\,|z, a)$ with respect to the total variation distance in (z, a) is replaced by Assumption 4.1-(b).

Theorem 4.12 *For any $f \in \mathbb{F}$, the stochastic kernel $p(\,\cdot\,|z, f(z))$ is positive Harris recurrent with unique invariant probability measure μ_f. Therefore, we have*

$$V(f, z) = \int_{\mathsf{Z}} c(z, f(z)) \mu_f(dz) =: \rho_f.$$

The Markov chain $\{Z_t\}_{t=0}^{\infty}$ *induced by* f *is geometrically ergodic; that is, there exist positive real numbers* R *and* $\kappa < 1$ *such that for every* $z \in \mathsf{Z}$

$$\sup_{f \in \mathbb{F}} \| p^t(\,\cdot\,|z, f(z)) - \mu_f \|_{TV} \leq R\kappa^t,$$

where R *and* κ *continuously depend on* α*. Finally, there exist* $f^* \in \mathbb{F}$ *and* $h^* \in B(\mathsf{Z})$ *such that the triplet* (h^*, f^*, ρ_{f^*}) *satisfies the average cost optimality equality (ACOE), i.e.,*

$$\begin{aligned} \rho_{f^*} + h^*(z) &= \min_{a \in \mathsf{A}} \biggl[c(z,a) + \int_{\mathsf{Z}} h^*(y) p(dy|z,a) \biggr] \\ &= c(z, f^*(z)) + \int_{\mathsf{Z}} h^*(y) p(dy|z, f^*(z)), \end{aligned}$$

and therefore,

$$\inf_{\pi \in \Pi} V(\pi, z) =: V^*(z) = \rho_{f^*}.$$

For each n, define the one-stage cost function $b_n : \mathsf{Z} \times \mathsf{A} \to [0, \infty)$ and the stochastic kernel q_n on Z given $\mathsf{Z} \times \mathsf{A}$ as

$$\begin{aligned} b_n(z,a) &:= \int c(x,a) \nu_{n,i_n(z)}(dx), \\ q_n(\,\cdot\,|z,a) &:= \int p(\,\cdot\,|x,a) \nu_{n,i_n(z)}(dx). \end{aligned}$$

Observe that c_n (the one-stage cost function of MDP_n) is the restriction of b_n to Z_n, and p_n (the stochastic kernel of MDP_n) is the pushforward of the measure q_n with respect to Q_n; that is, $c_n(z_{n,i}, a) = b_n(z_{n,i}, a)$ for all $i = 1, \ldots, k_n$ and $p_n(\,\cdot\,|z_{n,i}, a) = Q_n * q_n(\,\cdot\,|z_{n,i}, a)$.

For each n, let $\widehat{\mathrm{MDP}}_n$ be defined as a Markov decision process with the following components: Z is the state space, A is the action space, q_n is the transition probability, and c is the one-stage cost function; that is,

$$\widehat{\mathrm{MDP}}_n = \bigl(\mathsf{Z}, \mathsf{A}, q_n, c\bigr).$$

Similarly, let $\widetilde{\mathrm{MDP}}_n$ be defined as a Markov decision process with the following components: Z is the state space, A is the action space, q_n is the transition probability, and b_n is the one-stage cost function; that is,

$$\widetilde{\mathrm{MDP}}_n = \bigl(\mathsf{Z}, \mathsf{A}, q_n, b_n\bigr).$$

History spaces, policies, and cost functions are defined in a similar way as before. The models $\widehat{\mathrm{MDP}}_n$ and $\widetilde{\mathrm{MDP}}_n$ are used as transitions between the original MDP and MDP_n in a similar way as the operators F_n and $\hat{T}_n$ were used as transitions between T and T_n for the discounted cost. We note that a similar technique was used in the proof of [98, Theorem 2], which studied the approximation problem for finite-state-action MDPs. In [98] the one-stage cost function is first perturbed and then the transition probability is perturbed. We first perturb the transition probability and then the cost function. However, our proof method is otherwise quite different from that of [98, Theorem 2] since [98] assumes finite state and action spaces.

We note that a careful analysis of $\widetilde{\mathrm{MDP}}_n$ reveals that its Bellman optimality operator is essentially the operator $\hat{T}_n$. Hence, the value function of $\widetilde{\mathrm{MDP}}_n$ is the piecewise constant extension of the value function of MDP_n for the discounted cost. A similar conclusion will be made for the average cost in Lemma 4.13.

First, notice that if we define

$$\phi_n(z,a) := \int \phi(y,a)\nu_{n,i_n(z)}(dy),$$

$$\lambda_n := Q_n * \lambda,$$

then it is straightforward to prove that both $\widehat{\mathrm{MDP}}_n$ and $\widetilde{\mathrm{MDP}}_n$ satisfy Assumption 4.2-(d),(e) when ϕ is replaced by ϕ_n, and Assumption 4.2-(d),(e) is true for MDP_n when ϕ and λ are replaced by the restriction of ϕ_n to Z_n and λ_n, respectively. Hence, Theorem 4.12 holds (with the same R and κ) for $\widehat{\mathrm{MDP}}_n$, $\widetilde{\mathrm{MDP}}_n$, and MDP_n for all n. Therefore, we denote by $\hat{f}_n^*$, $\tilde{f}_n^*$, and f_n^* the optimal stationary policies of $\widehat{\mathrm{MDP}}_n$, $\widetilde{\mathrm{MDP}}_n$, and MDP_n with the corresponding average costs $\hat{\rho}^n_{\hat{f}_n^*}$, $\tilde{\rho}^n_{\tilde{f}_n^*}$, and $\rho^n_{f_n^*}$, respectively. Furthermore, we also write $\hat{\rho}^n_f$, $\tilde{\rho}^n_f$, and ρ^n_f to denote the average cost of any stationary policy f for $\widehat{\mathrm{MDP}}_n$, $\widetilde{\mathrm{MDP}}_n$, and MDP_n, respectively. The corresponding invariant probability measures are also denoted in a same manner, with μ replacing ρ.

The following lemma essentially says that MDP_n and $\widetilde{\mathrm{MDP}}_n$ are not very different.

Lemma 4.13 *The stationary policy given by the piecewise constant extension of the optimal policy f_n^* of MDP_n to Z (i.e., $f_n^* \circ Q_n$) is optimal for $\widetilde{\mathrm{MDP}}_n$ with the same cost function $\rho^n_{f_n^*}$. Hence, $\tilde{f}_n^* = f_n^* \circ Q_n$ and $\tilde{\rho}^n_{\tilde{f}_n^*} = \rho^n_{f_n^*}$.*

Proof Note that by Theorem 4.12 there exists $h_n^* \in B(\mathsf{Z}_n)$ such that the triplet $(h_n^*, f_n^*, \rho^n_{f_n^*})$ satisfies the ACOE for MDP_n. But it is straightforward to show that the triplet $(h_n^* \circ Q_n, f_n^* \circ Q_n, \rho^n_{f_n^*})$ satisfies the ACOE for $\widetilde{\mathrm{MDP}}_n$. By [55, Theorem 2.6 and Lemma 5.2], this implies that $f_n^* \circ Q_n$ is an optimal stationary policy for $\widetilde{\mathrm{MDP}}_n$ with cost function $\rho^n_{f_n^*}$. Hence $\tilde{f}_n^* = f_n^* \circ Q_n$ and $\tilde{\rho}^n_{\tilde{f}_n^*} = \rho^n_{f_n^*}$. □

The following theorem is the main result of this section. It states that if one applies the piecewise constant extension of the optimal stationary policy of MDP_n to the original MDP, the resulting cost function will converge to the value function of the original MDP.

Theorem 4.14 *The average cost of the optimal policy $\tilde{f}_n^*$ for $\widetilde{MDP}_n$, obtained by extending the optimal policy f_n^* of MDP_n to Z, converges to the value function $V^* = \rho_{f^*}$ of the original MDP, i.e.,*

$$\lim_{n\to\infty} |\rho_{\tilde{f}_n^*} - \rho_{f^*}| = 0.$$

Hence, to find a near optimal policy for the original MDP, it is sufficient to compute the optimal policy of MDP_n for sufficiently large n, and then extend this policy to the original state space.

To show the statement of Theorem 4.14 we will prove a series of auxiliary results.

Lemma 4.15 *For all $t \geq 1$ we have*

$$\lim_{n\to\infty} \sup_{(y,f)\in\mathsf{Z}\times\mathbb{F}} \left\| p^t(\,\cdot\,|y, f(y)) - q_n^t(\,\cdot\,|y, f(y)) \right\|_{TV} = 0.$$

Proof See Section 4.7. □

Using Lemma 4.15 we prove the following result.

Lemma 4.16 *We have $\sup_{f\in\mathbb{F}} |\hat{\rho}_f^n - \rho_f| \to 0$ as $n \to \infty$, where $\hat{\rho}_f^n$ is the cost function of the policy f for $\widehat{MDP}_n$ and ρ_f is the cost function of the policy f for the original MDP.*

Proof For any $t \geq 1$ and $y \in \mathsf{Z}$ we have

$$\begin{aligned}
\sup_{f\in\mathbb{F}} |\hat{\rho}_f^n - \rho_f| &= \sup_{f\in\mathbb{F}} \left| \int_{\mathsf{Z}} c(z, f(z))\hat{\mu}_f^n(dz) - \int_{\mathsf{Z}} c(z, f(z))\mu_f(dz) \right| \\
&\leq \sup_{f\in\mathbb{F}} \left| \int_{\mathsf{Z}} c(z, f(z))\hat{\mu}_f^n(dz) - \int_{\mathsf{Z}} c(z, f(z))q_n^t(dz|y, f(y)) \right| \\
&\quad + \sup_{f\in\mathbb{F}} \left| \int_{\mathsf{Z}} c(z, f(z))q_n^t(dz|y, f(y)) - \int_{\mathsf{Z}} c(z, f(z))p^t(dz|y, f(y)) \right| \\
&\quad + \sup_{f\in\mathbb{F}} \left| \int_{\mathsf{Z}} c(z, f(z))p^t(dz|y, f(y)) - \int_{\mathsf{Z}} c(z, f(z))\mu_f(dz) \right| \\
&\leq 2R\kappa^t \|c\| + \|c\| \sup_{(y,f)\in\mathsf{Z}\times\mathbb{F}} \left\| q_n^t(\,\cdot\,|y, f(y)) - p^t(\,\cdot\,|y, f(y)) \right\|_{TV} \text{ (by Theorem 4.12)},
\end{aligned}$$

where R and κ are the constants in Theorem 4.12. Then, the result follows from Lemma 4.15. □

The following theorem states that the value function of $\widehat{\text{MDP}}_n$ converges to the value function of the original MDP.

Lemma 4.17 *We have* $|\hat{\rho}^n_{\hat{f}^*_n} - \rho_{f^*}| \to 0$ *as* $n \to \infty$.

Proof Notice that

$$\begin{aligned} |\hat{\rho}^n_{\hat{f}^*_n} - \rho_{f^*}| &= \max(\hat{\rho}^n_{\hat{f}^*_n} - \rho_{f^*}, \rho_{f^*} - \hat{\rho}^n_{\hat{f}^*_n}) \\ &\le \max(\hat{\rho}^n_{f^*} - \rho_{f^*}, \rho_{\hat{f}^*_n} - \hat{\rho}^n_{\hat{f}^*_n}) \\ &\le \sup_f |\hat{\rho}^n_f - \rho_f|. \end{aligned}$$

Then, the result follows from Lemma 4.16. □

Lemma 4.18 *We have* $\sup_{f \in \mathbb{F}} |\tilde{\rho}^n_f - \hat{\rho}^n_f| \to 0$ *as* $n \to \infty$.

Proof It is straightforward to show that $b_n \to c$ uniformly. Since the probabilistic structure of $\widetilde{\text{MDP}}_n$ and $\widehat{\text{MDP}}_n$ are the same (i.e., $\hat{\mu}^n_f = \tilde{\mu}^n_f$ for all f), we have

$$\begin{aligned} \sup_{f \in \mathbb{F}} |\tilde{\rho}^n_f - \hat{\rho}^n_f| &= \sup_{f \in \mathbb{F}} \left| \int_{\mathsf{Z}} b_n(z, f(z)) \hat{\mu}^n_f(dz) - \int_{\mathsf{Z}} c(z, f(z)) \hat{\mu}^n_f(dz) \right| \\ &\le \sup_{f \in \mathbb{F}} \int_{\mathsf{Z}} |b_n(z, f(z)) - c(z, f(z))| \hat{\mu}^n_f(dz) \\ &\le \|b_n - c\|. \end{aligned}$$

This completes the proof. □

The next lemma states that the difference between the value functions of $\widetilde{\text{MDP}}_n$ and $\widehat{\text{MDP}}_n$ converges to zero.

Lemma 4.19 *We have* $|\tilde{\rho}^n_{\tilde{f}^*_n} - \hat{\rho}^n_{\hat{f}^*_n}| \to 0$ *as* $n \to \infty$.

Proof See the proof of Lemma 4.17. □

The following result states that if we apply the optimal policy of $\widetilde{\text{MDP}}_n$ to $\widehat{\text{MDP}}_n$, then the resulting cost converges to the value function of $\widehat{\text{MDP}}_n$.

Lemma 4.20 *We have* $|\hat{\rho}^n_{\tilde{f}^*_n} - \hat{\rho}^n_{\hat{f}^*_n}| \to 0$ *as* $n \to \infty$.

Proof Since $|\hat{\rho}^n_{\tilde{f}^*_n} - \hat{\rho}^n_{\hat{f}^*_n}| \le |\hat{\rho}^n_{\tilde{f}^*_n} - \tilde{\rho}^n_{\tilde{f}^*_n}| + |\tilde{\rho}^n_{\tilde{f}^*_n} - \hat{\rho}^n_{\hat{f}^*_n}|$, the result follows from Lemmas 4.18 and 4.19. □

Now, we are ready to prove the main result of this section.

Proof of Theorem 4.14. We have $|\rho_{\tilde{f}_n^*} - \rho_{f^*}| \leq |\rho_{\tilde{f}_n^*} - \hat{\rho}^n_{\tilde{f}_n^*}| + |\hat{\rho}^n_{\tilde{f}_n^*} - \hat{\rho}^n_{\hat{f}_n^*}| + |\hat{\rho}^n_{\hat{f}_n^*} - \rho^n_{f^*}|$. The result now follows from Lemmas 4.16, 4.20, and 4.17. □

4.3 Finite-State Approximation of Non-compact State MDPs

To tackle problem **(Q1)** for non-compact state MDPs with unbounded one-stage cost, we use the following strategy: (i) first, we define a sequence of compact-state MDPs to approximate the original MDP, (ii) we use Theorems 4.3 and 4.14 to approximate the compact-state MDPs by finite-state models, and (iii) we prove the convergence of the finite-state models to the original model. In fact, steps (ii) and (iii) will be accomplished simultaneously.

Assumption 4.3 *The state space* X *is* σ*-compact.*

Since X is σ-compact, there exists a nested sequence of compact sets $\{K_n\}$ such that $\mathsf{X} = \bigcup_{n=1}^{\infty} K_n$. Let $\{\vartheta_n\}$ be a sequence of probability measures such that for each $n \geq 1$, $\vartheta_n \in \mathcal{P}(K_n^c)$ and

$$\gamma_n := \int_{K_n^c} w(x)\vartheta_n(dx) < \infty, \tag{4.4}$$

$$\gamma = \sup_n \tau_n := \sup_n \max\left\{0, \sup_{(x,a)\in\mathsf{X}\times\mathsf{A}} \int_{K_n^c} \big(\gamma_n - w(y)\big)\; p(dy|x,a)\right\} < \infty. \tag{4.5}$$

For example, such probability measures can be constructed by choosing $x_n \in K_n^c$ such that $w(x_n) < \inf_{x\in K_n^c} w(x) + \frac{1}{n}$ and letting $\vartheta_n(\,\cdot\,) = \delta_{x_n}(\,\cdot\,)$.

Similar to the finite-state MDP construction in Section 4.2, we define a sequence of compact-state MDPs, denoted as c-MDP$_n$, to approximate the original model. To this end, for each n let $\mathsf{X}_n = K_n \cup \{\Delta_n\}$, where $\Delta_n \in K_n^c$ is a so-called pseudo-state. We define the transition probability p_n on X_n given $\mathsf{X}_n \times \mathsf{A}$ and the one-stage cost function $c_n : \mathsf{X}_n \times \mathsf{A} \to [0,\infty)$ by

$$p_n(\,\cdot\,|x,a) = \begin{cases} p\big(\,\cdot\, \cap K_n|x,a\big) + p\big(K_n^c|x,a\big)\delta_{\Delta_n}(\,\cdot\,), & \text{if } x \in K_n \\ \int_{K_n^c} \Big(p\big(\,\cdot\, \cap K_n|z,a\big) + p\big(K_n^c|z,a\big)\delta_{\Delta_n}(\,\cdot\,)\Big)\vartheta_n(dz), & \text{if } x = \Delta_n, \end{cases}$$

$$c_n(x,a) = \begin{cases} c(x,a), & \text{if } x \in K_n \\ \int_{K_n^c} c(z,a)\vartheta_n(dz), & \text{if } x = \Delta_n. \end{cases}$$

With these definitions, c-MDP$_n$ is defined as a Markov decision process with the components $\big(\mathsf{X}_n, \mathsf{A}, p_n, c_n\big)$; that is,

$$\text{c-MDP}_n = \big(\mathsf{X}_n, \mathsf{A}, p_n, c_n\big).$$

History spaces, policies, and cost functions are defined in a similar way as in the original model. Let Π_n, Φ_n, and $\mathbb{F}_n$ denote the set of all policies, randomized stationary policies, and deterministic stationary policies of c-MDP$_n$, respectively. For each policy $\pi \in \Pi_n$ and initial distribution $\mu \in \mathscr{P}(\mathsf{X}_n)$, we denote the discounted and average cost functions for c-MDP$_n$ by $J_n(\pi, \mu)$ and $V_n(\pi, \mu)$, respectively.

To obtain the main result of this section, we introduce another MDP, denoted by $\overline{\text{MDP}}_n$, with the components $(\mathsf{X}, \mathsf{A}, q_n, b_n)$, where

$$q_n(\,\cdot\,|x,a) = \begin{cases} p(\,\cdot\,|x,a), & \text{if } x \in K_n \\ \int_{K_n^c} p(\,\cdot\,|z,a)\vartheta_n(dz), & \text{if } x \in K_n^c, \end{cases}$$

$$b_n(x,a) = \begin{cases} c(x,a), & \text{if } x \in K_n \\ \int_{K_n^c} c(z,a)\vartheta_n(dz), & \text{if } x \in K_n^c. \end{cases}$$

Hence we have

$$\overline{\text{MDP}}_n = \big(\mathsf{X}, \mathsf{A}, q_n, b_n\big).$$

For each policy $\pi \in \Pi$ and initial distribution $\mu \in \mathscr{P}(\mathsf{X})$, we denote the discounted and average cost functions for $\overline{\text{MDP}}_n$ by $\bar{J}_n(\pi, \mu)$ and $\bar{V}_n(\pi, \mu)$, respectively.

4.3.1 Discounted Cost

The following conditions will be imposed for the discounted cost.

Assumption 4.4 *Suppose Assumption 4.3 and Assumption 3.3 hold for some M, $\alpha \in [1, \frac{1}{\beta})$, and $w : \mathsf{X} \to [1, \infty)$.*

The following result states that c-MDP$_n$ and $\overline{\text{MDP}}_n$ are equivalent for the discounted cost.

Lemma 4.21 *We have*

$$\bar{J}_n^*(x) = \begin{cases} J_n^*(x), & \text{if } x \in K_n \\ J_n^*(\Delta_n), & \text{if } x \in K_n^c, \end{cases} \tag{4.6}$$

where $\bar{J}_n^*$ is the discounted cost value function of $\overline{\text{MDP}}_n$ and J_n^* is the discounted cost value function of c-MDP$_n$, provided that there exist optimal deterministic stationary policies for $\overline{\text{MDP}}_n$ and c-MDP$_n$. Furthermore, if for any deterministic *stationary policy* $f \in \mathbb{F}_n$, *we let* $\bar{f}(x) = f(x)$ *on* K_n *and* $\bar{f}(x) = f(\Delta_n)$ *on* K_n^c, *then*

$$\bar{J}_n(\bar{f}, x) = \begin{cases} J_n(f, x), & \text{if } x \in K_n \\ J_n(f, \Delta_n), & \text{if } x \in K_n^c. \end{cases} \tag{4.7}$$

In particular, if the deterministic stationary policy $f_n^* \in \mathbb{F}_n$ *is optimal for c-MDP*$_n$, *then its extension* $\bar{f}_n^*$ *to* X *is also optimal for* $\overline{MDP}_n$.

Proof The proof of (4.7) is a consequence of the following facts: $b_n(x, a) = b_n(y, a)$ and $q_n(\cdot|x, a) = q_n(\cdot|y, a)$ for all $x, y \in K_n^c$ and $a \in \mathsf{A}$. In other words, K_n^c in $\overline{\text{MDP}}_n$ behaves like the pseudo state Δ_n in c-MDP$_n$ when $\bar{f}$ is applied to $\overline{\text{MDP}}_n$.

Let $\overline{\mathbb{F}}_n$ denote the set of all deterministic stationary policies in $\mathbb{F}$ which are obtained by extending policies in $\mathbb{F}_n$ to X. If we can prove that $\min_{f \in \mathbb{F}} \bar{J}_n(f, x) = \min_{f \in \overline{\mathbb{F}}_n} \bar{J}_n(f, x)$ for all $x \in \mathsf{X}$, then (4.6) follows from (4.7). Let $f \in \mathbb{F} \setminus \overline{\mathbb{F}}_n$. We have two cases: (i) $\bar{J}_n(f, z) = \bar{J}_n(f, y)$ for all $z, y \in K_n^c$ or (ii) there exists $z, y \in K_n^c$ such that $\bar{J}_n(f, z) < \bar{J}_n(f, y)$.

For the case (i), if we define the deterministic Markov policy π^0 as $\pi^0 = \{f_0, f, f, \ldots\}$, where $f_0(x) = f(z)$ on K_n^c for some fixed $z \in K_n^c$ and $f_0(x) = f(x)$ on K_n, then using the expression

$$\bar{J}_n(\pi^0, x) = b_n(x, f_0(x)) + \beta \int_{\mathsf{X}} \bar{J}_n(f, x') q_n(dx'|x, f_0(x)), \tag{4.8}$$

it is straightforward to show that $\bar{J}_n(\pi^0, x) = \bar{J}_n(f, x)$ on K_n and $\bar{J}_n(\pi^0, x) = \bar{J}_n(f, z)$ on K_n^c. Therefore, $\bar{J}_n(\pi^0, x) = \bar{J}_n(f, x)$ for all $x \in \mathsf{X}$ since $\bar{J}_n(f, x) = \bar{J}_n(f, z)$ for all $x \in K_n^c$. For all $t \geq 1$ define the deterministic Markov policy π^t as $\pi^t = \{f_0, \pi^{t-1}\}$. Analogously, one can prove that $\bar{J}_n(\pi^t, x) = \bar{J}_n(\pi^{t+1}, x)$ for all $x \in \mathsf{X}$. Since $\bar{J}_n(\pi^t, x) \to \bar{J}_n(f_0, x)$ as $t \to \infty$, we have $\bar{J}_n(f_0, x) = \bar{J}_n(f, x)$ for all $x \in \mathsf{X}$, where $f_0 \in \overline{\mathbb{F}}_n$.

For the second case, if we again consider the deterministic Markov policy $\pi^0 = \{f_0, f, f, \ldots\}$, then by (4.8) we have $\bar{J}_n(\pi^0, y) = \bar{J}_n(f, z) < \bar{J}_n(f, y)$. Since

$$\min_{f \in \mathbb{F}} \bar{J}_n(f, y) \leq \bar{J}_n(\pi^0, y),$$

this completes the proof. □

For each n, let us define w_n by letting $w_n(x) = w(x)$ on K_n and $w_n(x) = \int_{K_n^c} w(z)\vartheta_n(dz) =: \gamma_n$ on K_n^c. Hence, $w_n \in B(\mathsf{X})$ by (4.4).

Lemma 4.22 *For all* n *and* $x \in \mathsf{X}$, *the components of* $\overline{MDP}_n$ *satisfy the following:*

$$\sup_{a \in \mathsf{A}} b_n(x, a) \leq M w_n(x) \tag{4.9}$$

$$\sup_{a \in \mathsf{A}} \int_{\mathsf{X}} w_n(y) q_n(dy|x, a) \leq \alpha w_n(x) + \gamma, \tag{4.10}$$

where γ is the constant in (4.5).

Proof See Section 4.7. □

Note that if we define $c_{n,0}(x) = 1 + \sup_{a \in \mathsf{A}} b_n(x, a)$ and

$$c_{n,t}(x) = \sup_{a \in \mathsf{A}} \int_{\mathsf{X}} c_{n,t-1}(y) q_n(dy|x, a),$$

then by (4.9) and (4.10), and an induction argument, we obtain (see [69, p. 46])

$$c_{n,t}(x) \leq L w_n(x) \alpha^t + L\gamma \sum_{j=0}^{t-1} \alpha^j \text{ for all } x \in \mathsf{X}, \tag{4.11}$$

where $L = 1 + M$. Let $\beta_0 > \beta$ be such that $\alpha\beta_0 < 1$ and let $C_n : \mathsf{X} \to [1, \infty)$ be defined by

$$C_n(x) = \sum_{t=0}^{\infty} \beta_0^t c_{n,t}(x).$$

Then, for all $x \in \mathsf{X}$, by (4.11) we have

$$\begin{aligned} C_n(x) := \sum_{t=0}^{\infty} \beta_0^t c_{n,t}(x) &\leq \frac{L}{1 - \beta_0 \alpha} w_n(x) + \frac{L\beta_0}{(1 - \beta_0)(1 - \beta_0 \alpha)} \gamma \\ &:= L_1 w_n(x) + L_2. \end{aligned} \tag{4.12}$$

Hence $C_n \in B(\mathsf{X})$ as $w_n \in B(\mathsf{X})$. Moreover, for all $(x, a) \in \mathsf{X} \times \mathsf{A}$, C_n satisfies (see [69, p. 45])

$$\begin{aligned} \int_{\mathsf{X}} C_n(y) q_n(dy|x, a) &= \sum_{t=0}^{\infty} \beta_0^t \int_{\mathsf{X}} c_{n,t}(y) q_n(dy|x, a) \\ &\leq \sum_{t=0}^{\infty} \beta_0^t c_{n,t+1}(x) \\ &\leq \frac{1}{\beta_0} \sum_{t=0}^{\infty} \beta_0^t c_{n,t}(x) = \alpha_0 C_n(x), \end{aligned}$$

where $\alpha_0 := \frac{1}{\beta_0}$ and $\alpha_0 \beta < 1$ since $\beta_0 > \beta$. Therefore, for all $x \in \mathsf{X}$, components of $\overline{\text{MDP}}_n$ satisfy

$$\sup_{a \in \mathsf{A}} b_n(x,a) \leq C_n(x) \tag{4.13}$$

$$\sup_{a \in \mathsf{A}} \int_{\mathsf{X}} C_n(y) q_n(dy|x,a) \leq \alpha_0 C_n(x). \tag{4.14}$$

Since $C_n \in B(\mathsf{X})$, the Bellman optimality operator $\overline{T}_n$ of $\overline{\text{MDP}}_n$ maps $B^l(\mathsf{X})$ into $B^l(\mathsf{X})$ and is given by

$$\begin{aligned}\overline{T}_n u(x) &= \min_{a \in \mathsf{A}} \biggl[b_n(x,a) + \beta \int_{\mathsf{X}} u(y) q_n(dy|x,a) \biggr] \\ &= \begin{cases} \min_{a \in \mathsf{A}} \bigl[c(x,a) + \beta \int_{\mathsf{X}} u(y) p(dy|x,a) \bigr], & \text{if } x \in K_n \\ \min_{a \in \mathsf{A}} \int_{K_n^c} \bigl[c(z,a) + \beta \int_{\mathsf{X}} u(y) p(dy|z,a) \bigr] \vartheta_n(dz), & \text{if } x \in K_n^c. \end{cases}\end{aligned}$$

Then successive approximations to the discounted cost value function of $\overline{\text{MDP}}_n$ are given by $v_n^0 = 0$ and $v_n^{t+1} = \overline{T}_n v_n^t$ $(t \geq 1)$. Since $\alpha_0 \beta < 1$, it can be proved as in [69, Theorem 8.3.6, p. 47] and [69, (8.3.34), p. 52] that

$$v_n^t(x) \leq \bar{J}_n^*(x) \leq \frac{C_n(x)}{1-\sigma_0} \quad \text{for all } x, \tag{4.15}$$

$$\|v_n^t - \bar{J}_n^*\|_{C_n} \leq \frac{\sigma_0^t}{1-\sigma_0}, \tag{4.16}$$

where $\sigma_0 = \beta \alpha_0 < 1$.

Similar to v_n^t, let us define $v^0 = 0$ and $v^{t+1} = T v^t$, where $T : B_w^l(\mathsf{X}) \to B_w^l(\mathsf{X})$, the Bellman optimality operator for the original MDP is given by

$$Tu(x) = \min_{a \in \mathsf{A}} \biggl[c(x,a) + \beta \int_{\mathsf{X}} u(y) p(dy|x,a) \biggr].$$

Then, again by [69, Theorem 8.3.6, p. 47] and [69, (8.3.34), p. 52] we have

$$v^t(x) \leq J^*(x) \leq M \frac{w(x)}{1-\sigma} \quad \text{for all } x, \tag{4.17}$$

$$\|v^t - J^*\|_w \leq M \frac{\sigma^t}{1-\sigma}, \tag{4.18}$$

where $\sigma = \beta \alpha < 1$.

Lemma 4.23 *For any compact set $K \subset \mathsf{X}$, we have*

$$\lim_{n\to\infty} \sup_{x\in K} |v_n^t(x) - v^t(x)| = 0. \tag{4.19}$$

Proof We prove (4.19) by induction on t. For $t = 1$, the claim trivially holds since any compact set $K \subset \mathsf{X}$ is inside K_n for sufficiently large n, and therefore, $b_n = c$ on K for sufficiently large n (recall the $v_n^0 = v^0 = 0$). Assume the claim is true for $t \geq 1$. Fix any compact set K. Recall the definition of compact subsets K_ε of X in Lemma 3.17. By definition of q_n, b_n, and w_n, there exists $n_0 \geq 1$ such that for all $n \geq n_0$, $q_n = p$, $b_n = c$, and $w_n = w$ on K. With these observations, for each $n \geq n_0$ we have

$$\begin{aligned}
&\sup_{x\in K} |v_n^{t+1}(x) - v^{t+1}(x)| \\
&= \sup_{x\in K} \left| \min_{\mathsf{A}} \left[c(x,a) + \beta \int_{\mathsf{X}} v_n^t(y) p(dy|x,a) \right] - \min_{\mathsf{A}} \left[c(x,a) + \beta \int_{\mathsf{X}} v^t(y) p(dy|x,a) \right] \right| \\
&\leq \beta \sup_{(x,a)\in K\times\mathsf{A}} \left| \int_{\mathsf{X}} v_n^t(y) p(dy|x,a) - \int_{\mathsf{X}} v^t(y) p(dy|x,a) \right| \\
&= \beta \sup_{(x,a)\in K\times\mathsf{A}} \left| \int_{K_\varepsilon} \left(v_n^t(y) - v^t(y) \right) p(dy|x,a) + \int_{K_\varepsilon^c} \left(v_n^t(y) - v^t(y) \right) p(dy|x,a) \right| \\
&\leq \beta \left\{ \sup_{x\in K_\varepsilon} |v_n^t(x) - v^t(x)| + \sup_{(x,a)\in K\times\mathsf{A}} \left| \int_{K_\varepsilon^c} \left(v_n^t(y) - v^t(y) \right) p(dy|x,a) \right| \right\}
\end{aligned}$$

Note that we have $v^t \leq M \frac{w}{1-\sigma}$ by (4.17). Since $w_n \leq \gamma_{\max} w$, where $\gamma_{\max} := \max\{1, \gamma\}$, we also have $v_n^t \leq \frac{L_1 \gamma_{\max} w + L_2}{1-\sigma_0} \leq \frac{(L_1 \gamma_{\max} + L_2) w}{1-\sigma_0}$ by (4.12) and (4.15) (as $w \geq 1$). Let us define

$$R := \frac{L_1 \gamma_{\max} + L_2}{1-\sigma_0} + \frac{M}{1-\sigma}.$$

Then by Lemma 3.17 we have

$$\sup_{x\in K} |v_n^{t+1}(x) - v^{t+1}(x)| \leq \beta \sup_{x\in K_\varepsilon} |v_n^t(x) - v^t(x)| + \beta R \varepsilon.$$

Since the first term converges to zero as $n \to \infty$ by the induction hypothesis, and ε is arbitrary, the claim is true for $t + 1$. This completes the proof. □

The following theorem states that the discounted cost value function of $\overline{\text{MDP}}_n$ converges to the discounted cost value function of the original MDP uniformly on each compact set $K \subset \mathsf{X}$.

Theorem 4.24 *For any compact set $K \subset \mathsf{X}$ we have*

$$\lim_{n\to\infty} \sup_{x\in K} |\bar{J}_n^*(x) - J^*(x)| = 0. \tag{4.20}$$

Proof Fix any compact set $K \subset \mathsf{X}$. Since w is continuous and therefore bounded on K, it is sufficient to prove $\lim_{n\to\infty} \sup_{x\in K} \frac{|\bar{J}_n^*(x) - J^*(x)|}{w(x)} = 0$. Let n be chosen such that $K \subset K_n$, and so $w_n = w$ on K. Then we have

$$\begin{aligned}
&\sup_{x\in K} \frac{|\bar{J}_n^*(x) - J^*(x)|}{w(x)} \\
&\leq \sup_{x\in K} \frac{|\bar{J}_n^*(x) - v_n^t(x)|}{w(x)} + \sup_{x\in K} \frac{|v_n^t(x) - v^t(x)|}{w(x)} + \sup_{x\in K} \frac{|v^t(x) - J^*(x)|}{w(x)} \\
&\leq \sup_{x\in K} \frac{|\bar{J}_n^*(x) - v_n^t(x)|}{C_n(x)} \frac{C_n(x)}{w(x)} + \sup_{x\in K} \frac{|v_n^t(x) - v^t(x)|}{w(x)} + M\frac{\sigma^t}{1-\sigma} \quad \text{(by (4.18))} \\
&\leq \sup_{x\in K} \frac{|\bar{J}_n^*(x) - v_n^t(x)|}{C_n(x)} \frac{(L_1 w_n(x) + L_2)}{w(x)} + \sup_{x\in K} \frac{|v_n^t(x) - v^t(x)|}{w(x)} + \frac{M\sigma^t}{1-\sigma} \quad \text{(by (4.12))} \\
&\leq (L_1 + L_2) \sup_{x\in K} \frac{|\bar{J}_n^*(x) - v_n^t(x)|}{C_n(x)} + \sup_{x\in K} \frac{|v_n^t(x) - v^t(x)|}{w(x)} + \frac{M\sigma^t}{1-\sigma} (w_n = w \text{ on } K) \\
&\leq (L_1 + L_2) \frac{\sigma_0^t}{1-\sigma_0} + \sup_{x\in K} \frac{|v_n^t(x) - v^t(x)|}{w(x)} + \frac{M\sigma^t}{1-\sigma} \quad \text{(by (4.16))}.
\end{aligned}$$

Since $w \geq 1$ on X, $\sup_{x\in K} \frac{|v_n^t(x) - v^t(x)|}{w(x)} \to 0$ as $n \to \infty$ for all t by Lemma 4.23. Hence, the last expression can be made arbitrarily small. This completes the proof. □

In the remainder of this section, we use the above results and Theorem 4.3 to construct a near optimal policy for the original MDP. It is straightforward to check that for each n, c-MDP$_n$ satisfies the assumptions in Theorem 4.3. Let $\{\varepsilon_n\}$ be a sequence of positive real numbers such that $\lim_{n\to\infty} \varepsilon_n = 0$.

By Theorem 4.3, for each n there exists a deterministic stationary policy $f_n \in \mathbb{F}_n$, obtained from the finite-state approximations of c-MDP$_n$, such that

$$\sup_{x\in \mathsf{X}_n} |J_n(f_n, x) - J_n^*(x)| \leq \varepsilon_n,$$

where for each n, finite-state models are constructed replacing $(\mathsf{Z}, \mathsf{A}, p, c)$ with the components $(\mathsf{X}_n, \mathsf{A}, p_n, c_n)$ of c-MDP$_n$ in Section 4.2. By Lemma 4.21, for each $n \geq 1$ we also have

$$\sup_{x\in \mathsf{X}} |\bar{J}_n(f_n, x) - \bar{J}_n^*(x)| \leq \varepsilon_n, \tag{4.21}$$

where, with an abuse of notation, we also denote the extended (to X) policy by f_n. Let us define operators $\bar{R}_n : B_{C_n}(\mathsf{X}) \to B_{C_n}(\mathsf{X})$ and $R_n : B_w(\mathsf{X}) \to B_w(\mathsf{X})$ by

$$\bar{R}_n u(x) = \begin{cases} c(x, f_n(x)) + \beta \int_{\mathsf{X}} u(y) p(dy|x, f_n(x)), & \text{if } x \in K_n \\ \int_{K_n^c} \left[c(z, f_n(z)) + \beta \int_{\mathsf{X}} u(y) p(dy|z, f_n(z))\right] \vartheta_n(dz), & \text{if } x \in K_n^c, \end{cases}$$

$$R_n u(x) = c(x, f_n(x)) + \beta \int_{\mathsf{X}} u(y) p(dy|x, f_n(x)).$$

By [69, Remark 8.3.10, p. 54], $\bar{R}_n$ is a contraction operator with modulus σ_0 and R_n is a contraction operator with modulus σ. Furthermore, the fixed point of $\bar{R}_n$ is $\bar{J}_n(f_n, x)$ and the fixed point of R_n is $J(f_n, x)$. For each $n \geq 1$, let us define $\bar{u}_n^0 = u_n^0 = 0$ and $\bar{u}_n^{t+1} = \bar{R}_n \bar{u}_n^t$, $u_n^{t+1} = R_n u_n^t$ $(t \geq 1)$. One can prove that (see the proof of [69, Theorem 8.3.6])

$$\bar{u}_n^t(x) \leq \bar{J}_n(f_n, x) \leq \frac{C_n(x)}{1 - \sigma_0}$$

$$\|\bar{u}_n^t - \bar{J}_n(f_n, \cdot)\|_{C_n} \leq \frac{\sigma_0^t}{1 - \sigma_0}$$

$$u_n^t(x) \leq J(f_n, x) \leq M \frac{w(x)}{1 - \sigma}$$

$$\|u_n^t - J(f_n, \cdot)\|_w \leq M \frac{\sigma^t}{1 - \sigma}.$$

Lemma 4.25 *For any compact set $K \subset \mathsf{X}$, we have*

$$\lim_{n \to \infty} \sup_{x \in K} |\bar{u}_n^t(x) - u_n^t(x)| = 0.$$

Proof The lemma can be proved using the same arguments as in the proof of Lemma 4.23 and so we omit the details. □

Lemma 4.26 *For any compact set $K \subset \mathsf{X}$, we have*

$$\lim_{n \to \infty} \sup_{x \in K} |\bar{J}_n(f_n, x) - J(f_n, x)| = 0. \tag{4.22}$$

Indeed, this is true for all sequences of policies in $\mathbb{F}$.

Proof The lemma can be proved using the same arguments as in the proof of Theorem 4.24. □

The following theorem is the main result of this section which states that the true cost functions of the policies obtained from finite-state models converge to the value function of the original MDP. Hence, to obtain a near optimal policy for the original

MDP, it is sufficient to compute the optimal policy for the finite-state model that has sufficiently large number of grid points.

Theorem 4.27 *For any compact set $K \subset \mathsf{X}$, we have*

$$\lim_{n \to \infty} \sup_{x \in K} |J(f_n, x) - J^*(x)| = 0.$$

Therefore, $\lim_{n \to \infty} |J(f_n, x) - J^*(x)| = 0$ *for all* $x \in \mathsf{X}$.

Proof The result follows from (4.20), (4.21), and (4.22). □

4.3.2 Average Cost

Here we obtain approximation results, analogous to Theorems 4.24 and 4.27, for the average cost case. To do this, we impose some new assumptions on the components of the original MDP. These assumptions are the unbounded counterpart of Assumption 4.2.

Assumption 4.5 *Suppose Assumption 4.3 holds. Moreover, there exist $\lambda \in \mathcal{P}(\mathsf{X})$, $\alpha \in (0, 1)$, and a positive measurable function $\phi : \mathsf{X} \times \mathsf{A} \to (0, \infty)$ such that*

(a) For each $n \geq 1$, $\inf_{(x,a) \in K_n \times \mathsf{A}} \phi(x, a) > 0$.
(b) The stochastic kernel $p(\cdot\,|x, a)$ is continuous in (x, a) with respect to the w-norm.
(c) $\int_{\mathsf{X}} w(y) p(dy|x, a) \leq \alpha w(x) + \lambda(w)\phi(x, a)$.
(d) $p(D|x, a) \geq \lambda(D)\phi(x, a)$ for all $D \in \mathcal{B}(\mathsf{X})$.
(e) The weight function w is λ-integrable, i.e., $\lambda(w) < \infty$.

Throughout this section, it is assumed that Assumption 4.5 holds. Conditions (c) and (d) of Assumption 4.5 are the unbounded counterparts of conditions (e) and (d) in Assumption 4.2, respectively. Recall that condition (c) corresponds to the so-called "drift inequality" and condition (d) corresponds to the so-called "minorization" condition which guarantee the geometric ergodicity of Markov chains induced by stationary policies (see Section 3.3.2 and references therein). These assumptions are quite general for studying average cost problems with unbounded one-stage costs. In addition, they are well suited for the approximation problem in the sense that it can be shown that if the original problem satisfies these, then the reduced models constructed in the sequel satisfy similar conditions. As noted earlier in Section 4.2.2, we refer to [69, 96, 109] for alternative conditions that imply geometric ergodicity of Markov chains induced by stationary policies and which can replace conditions (c) and (d) in Assumption 4.5.

Note that although Assumption 4.5-(b) seems to be restrictive, it is weaker than the assumptions imposed in the literature for studying the approximation of average cost problems with unbounded cost (see [42]). Indeed, it is assumed in [42] that the transition probability p is Lipschitz continuous in (x, a) with respect to w-

norm. The reason for imposing such a strong condition on the transition probability is to obtain convergence rate for the approximation problem. Since we do not aim to provide rate of convergence result in this section, it is natural to impose continuity instead of Lipschitz continuity of the transition probability. However, it does not seem possible to replace continuity with respect to the w-norm by a weaker continuity notion. One reason is that with a weaker continuity assumption one cannot prove that the transition probability of c-MDP$_n$ is continuous with respect to the total variation distance, which is needed if one wants to use Theorem 4.14.

Analogous with Theorem 4.12, the following theorem is a consequence of [129, Theorems 3.3], [55, Lemma 3.4] (see also [69, Proposition 10.2.5, p. 126]), and [78, Theorem 3], which also holds with Assumption 4.5-(b) replaced by Assumption 3.3-(b).

Theorem 4.28 *For each $f \in \mathbb{F}$, the stochastic kernel $p(\,\cdot\,|x, f(x))$ has a unique invariant probability measure μ_f. Furthermore, w is μ_f-integrable, and therefore, $\rho_f := \int_{\mathsf{X}} c(x, f(x))\mu_f(dx) < \infty$. There exist positive real numbers R and $\kappa < 1$ such that*

$$\sup_{f \in \mathbb{F}} \|p^t(\,\cdot\,|x, f(x)) - \mu_f\|_w \leq R\kappa^t w(x) \tag{4.23}$$

for all $x \in \mathsf{X}$, where R and κ continuously depend on α, $\lambda(w)$, and $\inf_{f\in\mathbb{F}} \lambda(\phi(y, f(y)))$. Moreover, there exist $f^* \in \mathbb{F}$ and $h^* \in B_w(\mathsf{X})$ such that the triplet (h^*, f^*, ρ_{f^*}) satisfies the average cost optimality equality (ACOE), and therefore, for all $x \in \mathsf{X}$

$$\inf_{\pi\in\Pi} V(\pi, x) := V^*(x) = \rho_{f^*}.$$

Note that this theorem implies that for each $f \in \mathbb{F}$, the average cost is given by $V(f, x) = \int_{\mathsf{X}} c(y, f(y))\mu_f(dy)$ for all $x \in \mathsf{X}$ (instead of μ_f-a.e.).

Recall that V_n and $\bar{V}_n$ denote the average costs of c-MDP$_n$ and $\overline{\text{MDP}}_n$, respectively. The value functions for average cost are denoted analogously to the discounted cost case. Similar to Lemma 4.21, the following result states that c-MDP$_n$ and $\overline{\text{MDP}}_n$ are not too different for the average cost.

Lemma 4.29 *Suppose Theorem 4.28 holds for $\overline{\mathit{MDP}}_n$ and Theorem 4.12 holds for c-MDP$_n$. Then we have*

$$\bar{V}_n^*(x) = \begin{cases} V_n^*(x), & \text{if } x \in K_n \\ V_n^*(\Delta_n), & \text{if } x \in K_n^c. \end{cases} \tag{4.24}$$

Furthermore, if, for any deterministic stationary policy $f \in \mathbb{F}_n$, we define $\bar{f}(x) = f(x)$ on K_n and $\bar{f}(x) = f(\Delta_n)$ on K_n^c, then

$$\bar{V}_n(\bar{f}, x) = \begin{cases} V_n(f, x), & \text{if } x \in K_n \\ V_n(f, \Delta_n), & \text{if } x \in K_n^c. \end{cases} \tag{4.25}$$

In particular, if the deterministic stationary policy $f_n^* \in \mathbb{F}_n$ is optimal for c-MDP$_n$, then its extension $\bar{f}_n^*$ to X is also optimal for $\overline{\text{MDP}}_n$.

Proof Let the triplet $(h_n^*, f_n^*, \rho_{f_n^*}^n)$ satisfy the ACOE for c-MDP$_n$, so that f_n^* is an optimal policy and $\rho_{f_n^*}^n$ is the average cost value function for c-MDP$_n$. It is straightforward to show that the triplet $(\bar{h}_n^*, \bar{f}_n^*, \rho_{f_n^*}^n)$ satisfies the ACOE for $\overline{\text{MDP}}_n$, where

$$\bar{h}_n^*(x) = \begin{cases} h_n^*(x), & \text{if } x \in K_n \\ h_n^*(\Delta_n), & \text{if } x \in K_n^c, \end{cases}$$

and

$$\bar{f}_n^*(x) = \begin{cases} f_n^*(x), & \text{if } x \in K_n \\ f_n^*(\Delta_n), & \text{if } x \in K_n^c. \end{cases}$$

By [55, Theorem 2.6 and Lemma 5.2], this implies that $\bar{f}_n^*$ is an optimal stationary policy for $\overline{\text{MDP}}_n$ with cost function $\rho_{f_n^*}^n$. This completes the proof of the first part.

For the second part, let $f \in \mathbb{F}_n$ with an unique invariant probability measure $\mu_f \in \mathcal{P}(\mathsf{X}_n)$ and let $\bar{f} \in \mathbb{F}$ denote its extension to X with an unique invariant probability measure $\mu_{\bar{f}}$. Clearly, we have

$$\mu_f(\cdot) = \mu_{\bar{f}}(\cdot \cap K_n) + \mu_{\bar{f}}(K_n^c)\delta_{\Delta_n}(\cdot)$$

and therefore

$$\begin{aligned} \bar{V}_n(f, x) &= \int_{\mathsf{X}} b_n(x, \bar{f}(x))\mu_{\bar{f}}(dx) \\ &= \int_{K_n} c_n(x, \bar{f}(x))\mu_{\bar{f}}(dx) + \mu_{\bar{f}}(K_n^c)c_n(\Delta_n, \bar{f}(\Delta_n)) \\ &= \int_{\mathsf{X}_n} c_n(x, f(x))\mu_f(dx) \\ &= V_n(f, x). \end{aligned}$$

This completes the proof. □

By Lemma 4.29, in the remainder of this section we need only consider $\overline{\text{MDP}}_n$ in place of c-MDP$_n$. Later we will show that Theorem 4.28 holds for $\overline{\text{MDP}}_n$ for an n sufficiently large and that Theorem 4.12 holds for c-MDP$_n$ for all n.

Recall the definition of constants γ_n and τ_n from (4.4) and (4.5). For each $n \geq 1$, we define $\phi_n : \mathsf{X} \times \mathsf{A} \to (0, \infty)$ and $\varsigma_n \in \mathbb{R}$ as

$$\phi_n(x, a) := \begin{cases} \phi(x, a), & \text{if } x \in K_n \\ \int_{K_n^c} \phi(y, a)\vartheta_n(dy), & \text{if } x \in K_n^c, \end{cases}$$

$$\varsigma_n := \int_{K_n^c} w(y)\lambda(dy).$$

Since $\lambda(w) < \infty$ and τ_n can be made arbitrarily small by properly choosing ϑ_n, we assume, without loss of generality, the following.

Assumption 4.6 *The sequence of probability measures $\{\vartheta_n\}$ is chosen such that the following holds:*

$$\lim_{n\to\infty} (\tau_n + \varsigma_n) = 0. \tag{4.26}$$

Lemma 4.30 *For all n and $(x, a) \in \mathsf{X} \times \mathsf{A}$, the components of $\overline{MDP}_n$ satisfy the following:*

$$\sup_{a\in\mathsf{A}} b_n(x, a) \leq M w_n(x)$$

$$\int_{\mathsf{X}} w_n(y) q_n(dy|x, a) \leq \alpha_n w_n(x) + \lambda(w_n)\phi_n(x, a), \tag{4.27}$$

$$q_n(D|x, a) \geq \lambda(D)\phi_n(x, a) \text{ for all } D \in \mathscr{B}(\mathsf{X}),$$

where $\alpha_n := \alpha + \varsigma_n + \tau_n$.

Proof See Section 4.7. □

We note that by (4.26), there exists $n_0 \geq 1$ such that $\alpha_n < 1$ for $n \geq n_0$. Hence, for each $n \geq n_0$, Theorem 4.28 holds for $\overline{\text{MDP}}_n$ with w replaced by w_n for some $R_n > 0$, and $\kappa_n \in (0, 1)$, and we have $R_{\max} := \sup_{n \geq n_0} R_n < \infty$ and $\kappa_{\max} := \sup_{n \geq n_0} \kappa_n < 1$. Therefore, in the remainder of this section, it is assumed that $n \geq n_0$.

Lemma 4.31 *Let $g : \mathsf{X} \times \mathsf{A} \to \mathbb{R}$ be any measurable function such that $\sup_{a\in\mathsf{A}} |g(x, a)| \leq M_g w(x)$ for some $M_g \in \mathbb{R}$. Then, for all $t \geq 1$ and any compact set $K \subset \mathsf{X}$ we have*

$$\sup_{(y,f)\in K\times\mathbb{F}} \left| \int_{\mathsf{X}} g_n(x, f(x)) q_n^t(dx|y, f(y)) - \int_{\mathsf{X}} g(x, f(x)) p^t(dx|y, f(y)) \right| \to 0$$

as $n \to \infty$, where $g_n(x,a) = g(x,a)$ on $K_n \times \mathsf{A}$ and $g_n(x,a) = \int_{K_n^c} g(z,a)\vartheta_n(dz)$ on $K_n^c \times \mathsf{A}$.

Proof See Section 4.7. □

In the remainder of this section the above results are used to construct a near optimal policy for the original MDP. Let $\{\varepsilon_n\}$ be a sequence of positive real numbers converging to zero.

For each $f \in \mathbb{F}$, let μ_f^n denote the unique invariant probability measure of the transition kernel $q_n(\cdot\,|x, f(x))$ and let ρ_f^n denote the associated average cost; that is, $\rho_f^n := \bar{V}_n(f,x) = \int_{\mathsf{X}} b_n(y, f(y))\mu_f^n(dy)$ for all initial points $x \in \mathsf{X}$. Therefore, the value function of $\overline{\text{MDP}}_n$, denoted by $\bar{V}_n^*$, is given by $V_n^*(x) = \inf_{f \in \mathbb{F}} \rho_f^n$, i.e., it is constant on X.

Before making the connection with Theorem 4.14, we need the following result.

Lemma 4.32 *The transition probability p_n of c-MDP$_n$ is continuous in (x,a) with respect to the total variation distance.*

Proof See Section 4.7. □

Hence, we obtain that, for each $n \geq 1$, c-MDP$_n$ satisfies the assumption in Theorem 4.14 for

$$\tilde{\lambda}(\cdot) = \lambda(\cdot \cap K_n) + \lambda(K_n^c)\delta_{\Delta_n}(\cdot),$$

$$\tilde{\phi}(x,a) = \begin{cases} \phi(x,a), & \text{if } x \in K_n \\ \int_{K_n^c} \phi(y,a)\vartheta_n(dy), & \text{if } x = \Delta_n, \end{cases}$$

and some $\tilde{\alpha} \in (0,1)$, where the existence of $\tilde{\alpha}$ follows from Assumption 4.5-(a) and the fact that $\phi > 0$.

Consequently, there exists a deterministic stationary policy $f_n \in \mathbb{F}_n$, obtained from the finite-state approximations of c-MDP$_n$, such that

$$\sup_{x \in \mathsf{X}_n} |V_n(f_n, x) - V_n^*| \leq \varepsilon_n, \tag{4.28}$$

where the finite-state models are constructed replacing $(\mathsf{Z}, \mathsf{A}, p, c)$ with the components $(\mathsf{X}_n, \mathsf{A}, p_n, c_n)$ of c-MDP$_n$ in Section 4.2. By Lemma 4.29, we also have

$$|\rho_{f_n}^n - \bar{V}_n^*| \leq \varepsilon_n, \tag{4.29}$$

where, by an abuse of notation, we also denote the policy extended to X by f_n.

Lemma 4.33 *We have*

$$\sup_{f \in \mathbb{F}} |\rho_f^n - \rho_f| \to 0 \tag{4.30}$$

as $n \to \infty$.

Proof Fix any compact set $K \subset \mathsf{X}$. For any $t \geq 1$ and $y \in K$, we have

$$\sup_{f \in \mathbb{F}} |\rho_f^n - \rho_f| = \sup_{f \in \mathbb{F}} \biggl| \int_{\mathsf{X}} b_n(x, f(x)) \mu_f^n(dx) - \int_{\mathsf{X}} c(x, f(x)) \mu_f(dx) \biggr|$$

$$\leq \sup_{f \in \mathbb{F}} \biggl| \int_{\mathsf{X}} b_n(x, f(x)) \mu_f^n(dx) - \int_{\mathsf{X}} b_n(x, f(x)) q_n^t(dx|y, f(y)) \biggr|$$

$$+ \sup_{f \in \mathbb{F}} \biggl| \int_{\mathsf{X}} b_n(x, f(x)) q_n^t(dx|y, f(y)) - \int_{\mathsf{X}} c(x, f(x)) p^t(dx|y, f(y)) \biggr|$$

$$+ \sup_{f \in \mathbb{F}} \biggl| \int_{\mathsf{X}} c(x, f(x)) p^t(dx|y, f(y)) - \int_{\mathsf{X}} c(x, f(x)) \mu_f(dx) \biggr|$$

$$\leq M R_{\max} \kappa_{\max}^t w_n(y) + M R \kappa^t w(y)$$

$$+ \sup_{(y,f) \in K \times \mathbb{F}} \biggl| \int_{\mathsf{X}} b_n(x, f(x)) q_n^t(dx|y, f(y)) - \int_{\mathsf{X}} c(x, f(x)) p^t(dx|y, f(y)) \biggr|,$$

where the last inequality follows from Theorem 4.28. The result follows from Lemma 4.31. □

Theorem 4.34 *The value function of* $\overline{\text{MDP}}_n$ *converges to the value function of the original MDP, i.e.,*

$$|\bar{V}_n^* - V^*| \to 0$$

as $n \to \infty$.

Proof Since

$$|\bar{V}_n^* - V^*| = |\inf_{f \in \mathbb{F}} \rho_f^n - \inf_{f \in \mathbb{F}} \rho_f|$$
$$\leq \sup_{f \in \mathbb{F}} |\rho_f^n - \rho_f|,$$

the result follows from Lemma 4.33. □

The following is the main result of this section which states that the true average cost of the policies f_n obtained from finite-state approximations of c-MDP$_n$ converges to the average cost value function V^* of the original MDP.

Theorem 4.35 *We have*

$$|\rho_{f_n} - V^*| \to 0$$

as $n \to \infty$.

Proof We have

$$
\begin{aligned}
|\rho_{f_n} - V^*| &\leq |\rho_{f_n} - \rho^n_{f_n}| + |\rho^n_{f_n} - \bar{V}^*_n| + |\bar{V}^*_n - V^*| \\
&\leq \sup_{f \in \mathbb{F}} |\rho_f - \rho^n_f| + \varepsilon_n + |\bar{V}^*_n - V^*| \text{ (by (4.29))}.
\end{aligned}
$$

The result follows from Lemma 4.33 and Theorem 4.34. □

4.4 Discretization of the Action Space

For computing near optimal policies using well-known algorithms, such as value iteration, policy iteration, and Q-learning, the action space must be finite. In this section, using results from Chapter 3 we show that, as a preprocessing step, the action space can be taken to be finite with a sufficiently large number of points.

It was shown in Theorems 3.16 and 3.10 that any MDP with (infinite) compact action space and with bounded one-stage cost function can be well approximated by an MDP with finite-action space under assumptions that are satisfied by c-MDP$_n$ for both the discounted cost and the average cost cases. Recall the sequence of finite subsets $\{\Lambda_k\}$ of A from Section 3.3. We define c-MDP$_{n,k}$ as the Markov decision process having the components $\big\{\mathsf{X}_n, \Lambda_k, p_n, c_n\big\}$ and we let $\mathbb{F}_n(\Lambda_k)$ denote the set of all deterministic stationary policies for c-MDP$_{n,k}$. Note that $\mathbb{F}_n(\Lambda_k)$ is the set of policies in $\mathbb{F}_n$ taking values only in Λ_k. Therefore, in a sense, c-MDP$_{n,k}$ and c-MDP$_n$ can be viewed being the same MDP, where the former has constraints on the set of policies. For each n and k, by an abuse of notation, let f^*_n and $f^*_{n,k}$ denote the optimal stationary policies of c-MDP$_n$ and c-MDP$_{n,k}$, respectively, for both the discounted and average costs. Then Theorems 3.16 and 3.10 show that for all n, we have

$$
\begin{aligned}
\lim_{k \to \infty} J_n(f^*_{n,k}, x) &= J_n(f^*_n, x) := J^*_n(x) \\
\lim_{k \to \infty} V_n(f^*_{n,k}, x) &= V_n(f^*_n, x), := V^*_n(x)
\end{aligned}
$$

for all $x \in \mathsf{X}_n$. In other words, the discounted and average value functions of c-MDP$_{n,k}$ converge to the discounted and average value functions of c-MDP$_n$ as $k \to \infty$.

Theorem 4.36 *Suppose that Assumption 4.4 holds for the discounted cost and Assumption 4.5 holds for the average cost. Then for any $x \in \mathsf{X}$, there exists a subsequence $\{k_n\}$ such that*

$$
\lim_{n \to \infty} J(f^*_{n,k_n}, x) = J^*(x)
$$

$$\lim_{n\to\infty} V(f^*_{n,k_n}, x) = V^*(x),$$

where $f^*_{n,k_n} \in \mathbb{F}(\Lambda_{k_n})$ *is the extended optimal stationary policy of c-MDP*$_{n,k_n}$.

Proof Let us fix $x \in \mathsf{X}$. For n sufficiently large (so that $x \in K_n$), we choose k_n such that $|J_n(f^*_{n,k_n}, x) - J_n(f^*_n, x)| < 1/n$ (or $|V_n(f^*_{n,k_n}, x) - V_n(f^*_n, x)| < 1/n$ for the average cost). We note that if A is a compact subset of a finite dimensional Euclidean space, then by using Theorems 3.29 and 3.30 one can obtain an explicit expression for k_n in terms of n under further continuity conditions on c and p. By Lemmas 4.26 and 4.33, we have $|\bar{J}_n(f^*_{n,k_n}, x) - J(f^*_{n,k_n}, x)| \to 0$ and $|\bar{V}_n(f^*_{n,k_n}, x) - V(f^*_{n,k_n}, x)| \to 0$ as $n \to \infty$, where again by an abuse of notation, the policies extended to X are also denoted by f^*_{n,k_n}. Since $\bar{J}_n(f^*_{n,k_n}, x) = J_n(f^*_{n,k_n}, x)$ and $\bar{V}_n(f^*_{n,k_n}, x) = V_n(f^*_{n,k_n}, x)$, using Theorems 4.24 and 4.34 one can immediately obtain

$$\lim_{n\to\infty} J(f^*_{n,k_n}, x) = J^*(x)$$

$$\lim_{n\to\infty} V(f^*_{n,k_n}, x) = V^*(x).$$

□

Theorem 4.36 implies that before discretizing the state space to compute the near optimal policies, one can discretize, without loss of generality, the action space A in advance on a finite grid using sufficiently large number of grid points.

4.5 Rates of Convergence for Compact-State MDPs

In this section we consider problem **(Q2)** for MDPs with compact state space; that is, we derive an upper bound on the performance loss due to discretization in terms of the cardinality of the set Z_n (number of grid points). To do this, we will impose some new assumptions on the components of the MDP in addition to Assumptions 4.1 and 4.2. First, we present some definitions that are needed in the development.

For each $g \in C_b(\mathsf{Z})$, the Lipschitz seminorm of g is

$$\|g\|_{\mathrm{Lip}} := \sup_{z\neq y} \frac{|g(z) - g(y)|}{d_{\mathsf{Z}}(z, y)}.$$

If $\|g\|_{\mathrm{Lip}}$ is finite, then g is called Lipschitz continuous with Lipschitz constant $\|g\|_{\mathrm{Lip}}$. $\mathrm{Lip}(\mathsf{Z})$ denotes the set of all Lipschitz continuous functions on Z, i.e.,

$$\mathrm{Lip}(\mathsf{Z}) := \left\{g \in C_b(\mathsf{Z}) : \|g\|_{\mathrm{Lip}} < \infty\right\}$$

and $\mathrm{Lip}(\mathsf{Z}, K)$ denotes the set of all $g \in \mathrm{Lip}(\mathsf{Z})$ with $\|g\|_{\mathrm{Lip}} \leq K$. The *Wasserstein distance of order* 1 [130, p. 95] between two probability measures ζ and ξ over Z is defined as

$$W_1(\zeta, \xi) := \sup \biggl\{ \biggl| \int_{\mathsf{Z}} g d\zeta - \int_{\mathsf{Z}} g d\xi \biggr| : g \in \mathrm{Lip}(\mathsf{Z}, 1) \biggr\}.$$

We note that W_1 is also called the *Kantorovich-Rubinstein distance*. It is known that if Z is compact, then $W_1(\zeta, \xi) \leq \mathrm{diam}(\mathsf{Z}) \|\zeta - \xi\|_{TV}$ [130, Theorem 6.13]. For compact Z, the Wasserstein distance of order 1 is weaker than total variation distance. Furthermore, for compact Z, the Wasserstein distance of order 1 metrizes the weak topology on the set of probability measures $\mathscr{P}(\mathsf{Z})$ [130, Corollary 6.11], which also implies that convergence in this sense is weaker than setwise convergence.

In this section we impose the following supplementary assumptions in addition to Assumptions 4.1 and 4.2.

Assumption 4.7

(g) The one-stage cost function c satisfies $c(\,\cdot\,, a) \in \mathrm{Lip}(\mathsf{Z}, K_1)$ for all $a \in \mathsf{A}$ for some K_1.

(h) The stochastic kernel p satisfies $W_1\bigl(p(\,\cdot\,|z, a), p(\,\cdot\,|y, a)\bigr) \leq K_2 d_{\mathsf{Z}}(z, y)$ for all $a \in \mathsf{A}$ for some K_2.

(j) Z is an infinite compact subset of $\mathbb{R}^d$ for some $d \geq 1$, equipped with the Euclidean norm.

We note that Assumption 4.7-(j) implies the existence of a constant $\alpha > 0$ and finite subsets $\mathsf{Z}_n \subset \mathsf{Z}$ with cardinality n such that

$$\max_{z \in \mathsf{Z}} \min_{y \in \mathsf{Z}_n} d_{\mathsf{Z}}(z, y) \leq \alpha (1/n)^{1/d} \tag{4.31}$$

for all n, where d_{Z} is the Euclidean distance on Z. In the remainder of this section, we replace Z_n defined in Section 4.2 with Z_n satisfying (4.31) in order to derive *explicit* bounds on the approximation error in terms of the cardinality of Z_n.

4.5.1 Discounted Cost

Assumptions 4.1 and 4.7 are imposed throughout this section. Additionally, we assume that $K_2 \beta < 1$. The last assumption is the key to prove the next result which states that the value function J^* of the original MDP for the discounted cost is in $\mathrm{Lip}(\mathsf{Z})$. Although this result is known in the literature [75], we give a short proof for the sake of completeness using a simple application of the value iteration algorithm.

Theorem 4.37 *The value function J^* for the discounted cost is in $\mathrm{Lip}(\mathsf{Z}, K)$, where $K = K_1 \frac{1}{1 - \beta K_2}$.*

Proof Let $u \in \mathrm{Lip}(\mathsf{Z}, K)$ for some $K > 1$. Then $g = \frac{u}{K} \in \mathrm{Lip}(\mathsf{Z}, 1)$ and therefore for all $a \in \mathsf{A}$ and $z, y \in \mathsf{Z}$ we have

$$\left| \int_{\mathsf{Z}} u(x) p(dx|z,a) - \int_{\mathsf{Z}} u(x) p(dx|y,a) \right| = K \left| \int_{\mathsf{Z}} g(x) p(dx|z,a) - \int_{\mathsf{Z}} g(x) p(dx|y,a) \right|$$
$$\leq K W_1\big(p(\,\cdot\,|z,a), p(\,\cdot\,|y,a)\big) \leq K K_2 d_{\mathsf{Z}}(z,y),$$

by Assumption 4.7-(h). Hence, the Bellman optimality operator T of the MDP maps $u \in \mathrm{Lip}(\mathsf{Z}, K)$ to $Tu \in \mathrm{Lip}(\mathsf{Z}, K_1 + \beta K K_2)$, since, for all $z, y \in \mathsf{Z}$

$$|Tu(z) - Tu(y)|$$
$$\leq \max_{a \in \mathsf{A}} \left\{ |c(z,a) - c(y,a)| + \beta \left| \int_{\mathsf{Z}} u(x) p(dx|z,a) - \int_{\mathsf{Z}} u(x) p(dx|y,a) \right| \right\}$$
$$\leq K_1 d_{\mathsf{Z}}(z,y) + \beta K K_2 d_{\mathsf{Z}}(z,y) = \big(K_1 + \beta K K_2\big) d_{\mathsf{Z}}(z,y).$$

Now we apply T recursively to obtain the sequence $\{T^n u\}$ by letting $T^n u = T(T^{n-1}u)$, which converges to the value function J^* by the Banach fixed point theorem. Clearly, by induction we have for all $n \geq 1$

$$T^n u \in \mathrm{Lip}(\mathsf{Z}, K_n),$$

where $K_n = K_1 \sum_{i=0}^{n-1} (\beta K_2)^i + K(\beta K_2)^n$. If we choose $K < K_1$, then $K_n \leq K_{n+1}$ for all n and therefore, $K_n \uparrow K_1 \frac{1}{1-\beta K_2}$ since $K_2 \beta < 1$. Hence, $T^n u \in \mathrm{Lip}(\mathsf{Z}, K_1 \frac{1}{1-\beta K_2})$ for all n, and therefore, $J^* \in \mathrm{Lip}(\mathsf{Z}, K_1 \frac{1}{1-\beta K_2})$ since $\mathrm{Lip}(\mathsf{Z}, K_1 \frac{1}{1-\beta K_2})$ is closed with respect to the sup-norm. □

The following theorem is the main result of this section. Recall that the policy $\hat{f}_n \in \mathbb{F}$ is obtained by extending the optimal policy f_n^* of MDP_n to Z.

Theorem 4.38 *We have*

$$\|J(\hat{f}_n, \cdot) - J^*\| \leq \frac{\tau(\beta, K_2) K_1 \frac{1}{1-\beta K_2} + \frac{2K_1}{1-\beta}}{1-\beta} 2\alpha (1/n)^{1/d},$$

where $\tau(\beta, K_2) = (2+\beta)\beta K_2 + \frac{\beta^2 + 4\beta + 2}{(1-\beta)^2}$ *and* α *is the coefficient in (4.31).*

Proof To prove the theorem, we obtain upper bounds on the expressions derived in Section 4.2.1 in terms of the cardinality n of Z_n. The proof of Theorem 4.3 gives

$$\|J(\hat{f}_n, \cdot) - J^*\| \leq \frac{\|T_{\hat{f}_n} J^* - \hat{T}_{\hat{f}_n} J^*\| + (1+\beta)\|\hat{J}_n^* - J^*\|}{1-\beta}.$$

To prove the theorem we upper bound $\|T_{\hat{f}_n} J^* - \hat{T}_{\hat{f}_n} J^*\|$ and $\|\hat{J}_n^* - J^*\|$ in terms of n. For the first term we have

$$
\begin{aligned}
\|T_{\hat{f}_n} J^* - \hat{T}_{\hat{f}_n} J^*\| &= \sup_{z \in \mathsf{Z}} \left|T_{\hat{f}_n} J^*(z) - \hat{T}_{\hat{f}_n} J^*(z)\right| \\
&\leq \sup_{z \in \mathsf{Z}} \int \bigg| c(z, \hat{f}_n(z)) + \beta \int_{\mathsf{Z}} J^*(y) p(dy|z, \hat{f}_n(z)) - c(x, \hat{f}_n(x)) \\
&\qquad\qquad - \beta \int_{\mathsf{Z}} J^*(y) p(dy|x, \hat{f}_n(x)) \bigg| \nu_{n,i_n(z)}(dx) \\
&\leq \sup_{z \in \mathsf{Z}} \int \bigg(K_1 d_{\mathsf{Z}}(x, z) + \beta \bigg| \int_{\mathsf{Z}} J^*(y) p(dy|z, \hat{f}_n(z)) - \int_{\mathsf{Z}} J^*(y) p(dy|x, \hat{f}_n(z)) \bigg| \bigg) \nu_{n,i_n(z)}(dx) \\
&\qquad\qquad (\text{since } \hat{f}_n(x) = \hat{f}_n(z) \text{ for all } x \in \mathcal{S}_{n,i_n(z)}) \\
&\leq \sup_{z \in \mathsf{Z}} \int (K_1 + \beta \|J^*\|_{\mathrm{Lip}} K_2) d_{\mathsf{Z}}(x, z) \nu_{n,i_n(z)}(dx) \\
&\leq (K_1 + \beta \|J^*\|_{\mathrm{Lip}} K_2) \max_{i \in \{1,\ldots,n\}} \mathrm{diam}(\mathcal{S}_{n,i}) \\
&\leq (K_1 + \beta \|J^*\|_{\mathrm{Lip}} K_2) 2\alpha (1/n)^{1/d}. \qquad (4.32)
\end{aligned}
$$

For the second term, the proof of Theorem 4.10 gives

$$
\|\hat{J}_n^* - J^*\| \leq \frac{\|\hat{T}_n J^* - F_n J^*\| + (1 + \beta)\|J^* - u_n^*\|}{1 - \beta}.
$$

First consider $\|\hat{T}_n J^* - F_n J^*\|$. Define

$$
l(z, a) := c(z, a) + \beta \int_{\mathsf{X}} J^*(y) p(dy|z, a),
$$

so that

$$
J^*(z) = \min_{a \in \mathsf{A}} l(z, a).
$$

It is straightforward to show that $l(\,\cdot\,, a) \in \mathrm{Lip}(\mathsf{Z}, K_l)$ for all $a \in \mathsf{A}$, where $K_l = K_1 + \beta \|J^*\|_{\mathrm{Lip}} K_2$. By adapting the proof of Lemma 4.9 to the value function J^*, we obtain

$$
\begin{aligned}
\|\hat{T}_n J^* - F_n J^*\| &= \sup_{z \in \mathsf{Z}} \left| \min_{a \in \mathsf{A}} \int l(x, a) \nu_{n,i_n(z)}(dx) - \int \min_{a \in \mathsf{A}} l(x, a) \nu_{n,i_n(z)}(dx) \right| \\
&\leq \sup_{z \in \mathsf{Z}} \int \sup_{y \in \mathcal{S}_{n,i_n(z)}} \left| l(y, a_i) - J^*(y) \right| \nu_{n,i_n(z)}(dy)
\end{aligned}
$$

$$\leq \max_{i\in\{1,\dots,n\}} \int \sup_{y\in\mathcal{S}_{n,i}} \big\{|l(y,a_i) - l(z_i,a_i)| + |J^*(z_i) - J^*(y)|\big\} \nu_{n,i}(dy)$$

$$\leq \max_{i\in\{1,\dots,n\}} \int \sup_{y\in\mathcal{S}_{n,in}} \big\{K_l d_{\mathsf{Z}}(y,z_i) + \|J^*\|_{\mathrm{Lip}} d_{\mathsf{Z}}(z_i,y)\big\} \nu_{n,i}(dy)$$

$$\leq (K_l + \|J^*\|_{\mathrm{Lip}}) \max_{i\in\{1,\dots,n\}} \mathrm{diam}(\mathcal{S}_{n,i})$$

$$\leq (K_l + \|J^*\|_{\mathrm{Lip}}) 2\alpha(1/n)^{1/d}. \tag{4.33}$$

For the expression $\|J^* - u_n^*\|$, by Lemma 4.8 we have

$$\|u_n^* - J^*\| \leq \frac{2}{1-\beta} \inf_{r\in\mathsf{Z}^{k_n}} \|J^* - \Phi_r\|,$$

where $\Phi_r(z) = \Sigma_{i=1}^{k_n} r_i 1_{S_{n,i}}(z)$, $r = (r_1,\dots,r_{k_n})$. Since $\|J^*\|_{\mathrm{Lip}} < \infty$, we have $\inf_{r\in\mathsf{Z}^{k_n}} \|J^* - \Phi_r\| \leq \|J^*\|_{\mathrm{Lip}} \max_{i\in\{1,\dots,n\}} \mathrm{diam}(\mathcal{S}_{n,i}) \leq \|J^*\|_{\mathrm{Lip}} 2\alpha(1/n)^{1/d}$. Hence

$$\|u_n^* - J^*\| \leq \frac{2}{1-\beta} \|J^*\|_{\mathrm{Lip}} 2\alpha(1/n)^{1/d}. \tag{4.34}$$

Hence, by (4.33) and (4.34) we obtain

$$\|\hat{J}_n^* - J^*\| \leq \bigg((\beta K_2 + \frac{\beta+3}{(1-\beta)^2}) \|J^*\|_{\mathrm{Lip}} + \frac{K_1}{1-\beta} \bigg) 2\alpha(1/n)^{1/d}. \tag{4.35}$$

Then, the result follows from (4.32) and (4.35), and the fact $\|J^*\|_{\mathrm{Lip}} \leq K_1 \frac{1}{1-\beta K_2}$.
□

Remark 4.39 It is important to point out that if we replace Assumption 4.7-(h) with the uniform Lipschitz continuity of $p(\,\cdot\,|z,a)$ in z with respect to total variation distance, then Theorem 4.38 remains valid (with possibly different constants in front of the term $(1/n)^{1/d}$). However, in this case, we do not need the assumption $K_2\beta < 1$.

Remark 4.40 For the average cost case, instead of assuming from the outset the uniform Lipschitz continuity of c and p in the z variable, we first derive a rate of convergence result in terms of the moduli of continuity ω_c and ω_p of $c(z,a)$ and $p(\,\cdot\,|z,a)$ in the z variable, where the total variation distance is used to define ω_p. Then, we show that an explicit rate of convergence result can be given if we impose some structural assumptions on ω_c and ω_p such as linearity, which corresponds to the uniform Lipschitz continuity of $c(z,a)$ and $p(\,\cdot\,|z,a)$ in z. However, this is not the right approach for the discounted cost case as the modulus of continuity ω_p is calculated using the Wasserstein distance of order 1. Indeed, to obtain a similar result as in the average cost case, we must relate ω_c and ω_p to the modulus of continuity ω_{J^*} of the value function J^*. This can be established if ω_c and ω_p are affine functions (i.e., $\omega_c(r) = K_1 r + L_1$ and $\omega_p(r) = K_2 r + L_2$) using the dual formulation of the Wasserstein distance of order 1 [130, Theorem 5.10]:

$$W_1(\mu, \nu) = \sup_{\substack{(\psi,\varphi) \in C_b(\mathsf{Z}) \times C_b(\mathsf{Z}) \\ \psi(x) - \varphi(y) \leq d_{\mathsf{Z}}(x,y)}} \left| \int_{\mathsf{Z}} \psi(z)\mu(dz) - \int_{\mathsf{Z}} \varphi(z)\nu(dz) \right|.$$

However, in this situation we can explicitly compute the convergence rate only if $L_1 = L_2 = 0$ which is the uniform Lipschitz continuity case.

4.5.2 Average Cost

In this section, we suppose that Assumptions 4.2 and 4.7-(j) hold. We define the modulus of continuity functions in the z variable for $c(z, a)$ and $p(\cdot|z, a)$ as

$$\omega_c(r) := \sup_{a \in \mathsf{A}} \sup_{z,y \in \mathsf{Z}: d_{\mathsf{Z}}(z,y) \leq r} |c(z, a) - c(y, a)|$$

$$\omega_p(r) := \sup_{a \in \mathsf{A}} \sup_{z,y \in \mathsf{Z}: d_{\mathsf{Z}}(z,y) \leq r} \|p(\cdot|z, a) - p(\cdot|y, a)\|_{TV}.$$

Since $c(z, a)$ and $p(\cdot|z, a)$ are uniformly continuous, we have $\lim_{t \to 0} \omega_c(t) = 0$ and $\lim_{r \to 0} \omega_p(r) = 0$. Note that when ω_c and ω_p are linear, $c(z, a)$ and $p(\cdot|z, a)$ are uniformly Lipschitz in z. In the remainder of this section, we first derive a rate of convergence result in terms of ω_c and ω_c. Then, we explicitly compute the convergence rate for the Lipschitz case as a corollary of this result.

To obtain convergence rates for the average cost, we first prove a rate of convergence result for Lemma 4.15. To this end, for each $n \geq 1$, let $d_n := 2\alpha(1/n)^{1/d}$, where α is the coefficient in (4.31).

Lemma 4.41 *For all $t \geq 1$, we have*

$$\sup_{(y,f) \in \mathsf{Z} \times \mathbb{F}} \|p^t(\cdot|y, f(y)) - q_n^t(\cdot|y, f(y))\|_{TV} \leq t\omega_p(d_n).$$

Proof See Section 4.7. □

The following theorem is the main result of this section. Recall that the optimal policy $\tilde{f}_n^*$ for $\widetilde{\text{MDP}}_n$ is obtained by extending the optimal policy f_n^* for MDP_n to Z, and R and κ are the constants in Theorem 4.12.

Theorem 4.42 *For all $t \geq 1$, we have*

$$|\rho_{\tilde{f}_n^*} - \rho_{f^*}| \leq 4\|c\|R\kappa^t + 2\omega_c(d_n) + 2\|c\|t\omega_p(d_n).$$

Proof The proof of Theorem 4.14 gives

$$|\rho_{\tilde{f}_n^*} - \rho_{f^*}| \leq |\rho_{\tilde{f}_n^*} - \hat{\rho}^n_{\tilde{f}_n^*}| + |\hat{\rho}^n_{\tilde{f}_n^*} - \hat{\rho}^n_{\hat{f}_n^*}| + |\hat{\rho}^n_{\hat{f}_n^*} - \rho_{f^*}|.$$

Hence, to prove the theorem we obtain an upper bounds on the three terms in the sum. Consider the first term (recall the proof of Lemma 4.16)

$$\begin{aligned} |\rho_{\tilde{f}_n^*} - \hat{\rho}^n_{\tilde{f}_n^*}| &\leq \sup_{f \in \mathbb{F}} |\hat{\rho}^n_f - \rho_f| \\ &\leq 2R\kappa^t \|c\| + \|c\| \sup_{(y,f) \in \mathsf{Z} \times \mathbb{F}} \|q_n^t(\,\cdot\,|y, f(y)) - p^t(\,\cdot\,|y, f(y))\|_{TV} \\ &\leq 2R\kappa^t \|c\| + \|c\| t \omega_p(d_n) \text{ (by Lemma 4.41).} \end{aligned} \tag{4.36}$$

For the second term, the proof of Lemma 4.20 gives

$$\begin{aligned} |\hat{\rho}^n_{\tilde{f}_n^*} - \hat{\rho}^n_{\hat{f}_n^*}| &\leq |\hat{\rho}^n_{\tilde{f}_n^*} - \tilde{\rho}^n_{\tilde{f}_n^*}| + |\tilde{\rho}^n_{\tilde{f}_n^*} - \hat{\rho}^n_{\hat{f}_n^*}| \\ &\leq \sup_{f \in \mathbb{F}} |\hat{\rho}^n_f - \tilde{\rho}^n_f| + |\inf_{f \in \mathbb{F}} \tilde{\rho}^n_f - \inf_{f \in \mathbb{F}} \hat{\rho}^n_f| \\ &\leq 2 \sup_{f \in \mathbb{F}} |\hat{\rho}^n_f - \tilde{\rho}^n_f| \\ &\leq 2\|b_n - c\| \text{ (see the proof of Lemma 4.18)} \\ &\leq 2 \sup_{(z,a) \in \mathsf{Z} \times \mathsf{A}} \int |c(x,a) - c(z,a)| \nu_{n,i_n(z)}(dx) \\ &\leq 2\omega_c(d_n). \end{aligned} \tag{4.37}$$

For the last term, we have

$$\begin{aligned} |\hat{\rho}^n_{\hat{f}_n^*} - \rho_{f^*}| &= |\inf_{f \in \mathbb{F}} \hat{\rho}^n_f - \inf_{f \in \mathbb{F}} \rho_f| \leq \sup_{f \in \mathbb{F}} |\hat{\rho}^n_f - \rho_f| \\ &\leq 2R\kappa^t \|c\| + \|c\| t \omega_p(d_n) \text{ (by (4.36)).} \end{aligned} \tag{4.38}$$

Combining (4.36), (4.37), and (4.38) implies the result. □

To explicitly calculate a convergence rate, we need to impose some structural assumptions on ω_c and ω_p. One such assumption is linearity, which corresponds to the uniform Lipschitz continuity of $c(z, a)$ and $p(\,\cdot\,|z, a)$ in z. This means that $\omega_c(r) = K_1 r$ and $\omega_p(r) = K_2 r$, or equivalently, $|c(z, a) - c(y, a)| \leq K_1 d_{\mathsf{Z}}(z, y)$ and $\|p(\,\cdot\,|z, a) - p(\,\cdot\,|y, a)\| \leq K_2 d_{\mathsf{Z}}(z, y)$ for all $z, y \in \mathsf{Z}$ and $a \in \mathsf{Z}$. In this case, by Theorem 4.42, for all $t \geq 1$ we have

$$|\rho_{\tilde{f}_n^*} - \rho_{f^*}| \leq 4\|c\| R\kappa^t + 4K_1 \alpha (1/n)^{1/d} + 4\|c\| K_2 \alpha (1/n)^{1/d} t. \tag{4.39}$$

To obtain a proper rate of convergence result (i.e., an upper bound that only depends on n) the dependence on t should be eliminated. This can be done by (approximately) minimizing the upper bound in (4.39) with respect to t for each n as it is done in Section 3.5.2. Define the constants $I_1 := 4\|c\|R$, $I_2 := 4K_1\alpha$, and $I_3 := 4\|c\|K_2\alpha$. Then the upper bound in (4.39) becomes

$$I_1\kappa^t + I_2(1/n)^{1/d} + I_3(1/n)^{1/d}t. \tag{4.40}$$

It is straightforward to compute that

$$t'(n) := \ln\left(\frac{n^{1/d}}{I_4}\right)\frac{1}{\ln(\frac{1}{\kappa})}$$

is the zero of the derivative of the convex term in (4.40), where $I_4 := \frac{I_3}{I_1 \ln(\frac{1}{\kappa})}$. Letting $t = \lceil t'(n) \rceil$ in (4.40), we obtain the following result.

Corollary 4.43 *Suppose that $c(z,a)$ and $p(\cdot\,|z,a)$ are uniformly Lipschitz continuous in z in addition to the assumptions imposed at the beginning of this section. Then, we have*

$$|\rho_{\tilde{f}_n^*} - \rho_{f^*}| \leq (I_1 I_4 + I_2)(1/n)^{1/d} + \frac{I_3}{\ln(1/\kappa)}(1/n)^{1/d}\ln\left(\frac{n^{1/d}}{I_4}\right).$$

4.5.3 Order Optimality

The following example demonstrates that the order of the convergence rates in Theorem 4.38 and Corollary 4.43 cannot be better than $O((\frac{1}{n})^{\frac{1}{d}})$. More precisely, we exhibit a simple standard example where we can lower bound the performance loss by $L(1/n)^{1/d}$, for some positive constant L. A similar result was obtained in Section 3.5.3 for the case of quantization of action space. Therefore, when both state and action spaces are quantized, then the resulting construction is order optimal in the above sense as the approximation error, in this case, is bounded by the sum of the approximation errors in quantization of state space and quantization of action space.

Recall that $h(\cdot)$ and $h(\cdot\,|\,\cdot)$ denote differential and conditional differential entropies, respectively [33].

Consider the additive-noise system

$$Z_{t+1} = F(Z_t, A_t) + V_t, \; t = 0, 1, 2, \ldots,$$

where $Z_t, A_t, V_t \in \mathbb{R}^d$. We assume that

$$\sup_{(z,a)\in\mathbb{R}^d\times\mathbb{R}^d} \frac{\|F(z,a)\|}{\|z\|+\|a\|} < 1/2.$$

The noise process $\{V_t\}$ is a sequence of i.i.d. random vectors whose common distribution has density g supported on some compact subset V of $\mathbb{R}^d$. We choose V such that $\mathsf{Z} = \mathsf{A}$ can be taken to be compact subsets of $\mathbb{R}^d$. For simplicity suppose that the initial distribution μ has the same density g. We also assume that the differential entropy $h(g) := -\int_{\mathsf{Z}} g(z)\log g(z)dz$ is finite. Let the cost function be $c(z,a) := \|z-a\|$. Clearly, the optimal stationary policy f^* is induced by the identity $f^*(z) = z$, having the optimal cost $J(f^*,\mu) = 0$ and $V(f^*,\mu) = 0$. Let $\hat{f}_n$ be the piecewise constant extension of the optimal policy f_n^* of the MDP$_n$ to the set Z. Fix $n \geq 1$ and define $D_t := \mathbb{E}_{\mu}^{\hat{f}_n}\big[c(Z_t,A_t)\big]$ for all t. Then, since $A_t = \hat{f}_n(Z_t)$ can take at most n values in A, by the Shannon lower bound (SLB) [145, p. 12] we have for $t \geq 1$

$$\begin{aligned}
\log n \geq R(D_t) &\geq h(Z_t) + \theta(D_t) \\
&= h(F(Z_{t-1},A_{t-1}) + V_{t-1}) + \theta(D_t) \\
&\geq h(F(Z_{t-1},A_{t-1}) + V_{t-1}|Z_{t-1},A_{t-1}) + \theta(D_t) \\
&= h(V_{t-1}) + \theta(D_t),
\end{aligned} \tag{4.41}$$

where

$$\theta(D_t) = -d + \log\biggl(\frac{1}{dV_d\Gamma(d)}\Bigl(\frac{d}{D_t}\Bigr)^d\biggr),$$

$R(D_t)$ is the rate-distortion function of Z_t, V_d is the volume of the unit sphere $S_d = \{z : \|z\| \leq 1\}$, and Γ is the gamma function. Here, (4.41) follows from the independence of V_{t-1} and the pair (Z_{t-1},A_{t-1}). Note that $h(V_{t-1}) = h(g)$ for all t. Expressing D_t, we obtain $D_t \geq L(1/n)^{1/d}$, where

$$L := \frac{d}{2}\biggl(\frac{2^{h(g)}}{dV_d\Gamma(d)}\biggr)^{1/d}.$$

Since we have obtained stage-wise error bounds, these give

$$|J(f^*,\mu) - J(\hat{f}_n,\mu)| \geq \frac{L}{1-\beta}(1/n)^{1/d}$$

and

$$|V(f^*,\mu) - V(\hat{f}_n,\mu)| \geq L(1/n)^{1/d}.$$

4.6 Numerical Examples

In this section, we consider two examples, the additive noise model and the fisheries management problem, in order to illustrate our results numerically. Since computing true costs of the policies obtained from the finite models appears intractable, we only compute the value functions of the finite models and illustrate their converge to the value function of the original MDP as $n \to \infty$.

We note that all results in this chapter apply with straightforward modifications for the case of maximizing reward instead of minimizing cost.

4.6.1 Additive Noise System

In this example, the system dynamics are given by

$$X_{t+1} = F(X_t, A_t) + V_t, \ t = 0, 1, 2, \dots$$

where $X_t, A_t, V_t \in \mathbb{R}$ and $\mathsf{X} = \mathbb{R}$. The noise process $\{V_t\}$ is a sequence of $\mathbb{R}$-valued i.i.d. random variables with common density g. Hence, the transition probability $p(\,\cdot\,|x, a)$ is given by

$$p(D|x, a) = \int_D g(v - F(x, a))m(dv) \ \text{ for all } D \in \mathscr{B}(\mathbb{R}),$$

where m is the Lebesgue measure. The cost function is $c(x, a) = (x - a)^2$, the action space is $\mathsf{A} = [-L, L]$ for some $L > 0$, and the cost to be minimized is the discounted cost.

We assume that (i) g is a Gaussian probability density function with zero mean and variance σ^2, (ii) $\sup_{a \in \mathsf{A}} |F(x, a)|^2 \leq k_1 x^2 + k_2$ for some $k_1, k_2 \in \mathbb{R}_+$, (iii) $\beta < 1/\alpha$ for some $\alpha \geq k_1$, and (iv) F is continuous. Hence, Assumption 4.4 holds for this model with $w(x) = k + x^2$ and $M = 4\big(\frac{L^2}{k} + x^2\big)$, for some $k \in \mathbb{R}_+$.

For the numerical results, we use the parameters $F(x, a) = x + a$, $\beta = 0.3$, $L = 0.5$, and $\sigma = 0.1$.

We selected a sequence $\big\{[-l_n, l_n]\big\}_{n=1}^{15}$ of nested closed intervals, where $l_n = 0.5 + 0.25n$, to "approximate" $\mathbb{R}$. Each interval is uniformly discretized using $\lceil 2k_{\lceil \frac{n}{3} \rceil} l_n \rceil$ grid points, where $k_m = 5m$ for $m = 1, \dots, 5$ and $\lceil q \rceil$ denotes the smallest integer greater than or equal to $q \in \mathbb{R}$. Therefore, the discretization is gradually refined. For each n, the finite-state space is given by $\{x_{n,i}\}_{i=1}^{k_n} \cup \{\Delta_n\}$, where $\{x_{n,i}\}_{i=1}^{k_n}$ are the representation points in the uniform quantization of the closed interval $[-l_n, l_n]$ and Δ_n is a pseudo state. We also uniformly discretize the action space $\mathsf{A} = [-0.5, 0.5]$ using $2k_{\lceil \frac{n}{3} \rceil}$ grid points. For each n, the finite-state models are constructed as in Section 4.2 by replacing Z with $[-l_n, l_n]$ and by setting $\nu_n(\,\cdot\,) = \frac{1}{2} m_n(\,\cdot\,) + \frac{1}{2} \delta_{\Delta_n}(\,\cdot\,)$, where m_n is the Lebesgue measure normalized over $[-l_n, l_n]$.

The value iteration algorithm [68] is used to compute the value functions of the finite models. Figure 4.1 displays the graph of these value functions corresponding to the different values for the number of grid points, when the initial state is $x = 0.7$. The figure illustrates that the convergence of value functions of the finite models converge to the value function of the original model.

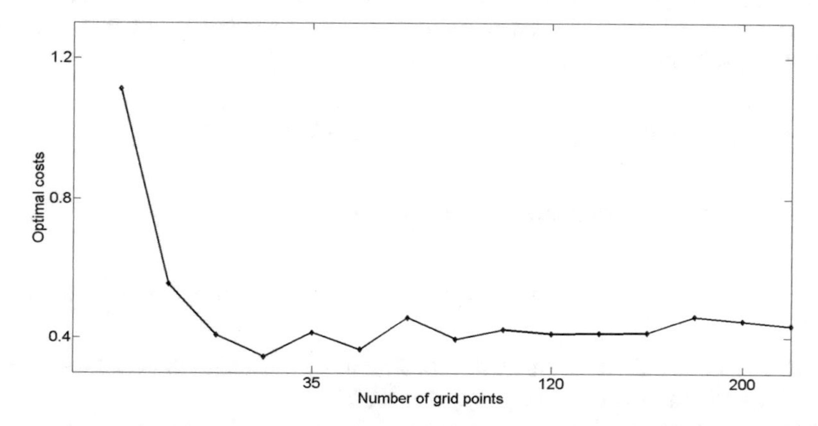

Fig. 4.1 Optimal costs of the finite models when the initial state is $x = 0.7$

4.6.2 *Fisheries Management Problem*

In this example we consider the population growth model, called a Ricker model (see [68, Section 1.3]), given by

$$X_{t+1} = \theta_1 A_t \exp\{-\theta_2 A_t + V_t\},\ t = 0, 1, 2, \dots \tag{4.42}$$

where $\theta_1, \theta_2 \in \mathbb{R}_+$, X_t is the fish population size in season t, and A_t is the population to be left for spawning for the next season, or in other words, $X_t - A_t$ is the amount of fish captured in the season t. The one-stage "reward" function is $u(x - a)$, where u is some utility function. In this model, the goal is to maximize the average reward.

The state and action spaces are $\mathsf{X} = \mathsf{A} = [\kappa_{\min}, \kappa_{\max}]$, for some $\kappa_{\min}, \kappa_{\max} \in \mathbb{R}_+$. Since the population left for spawning cannot be greater than the total population, for each $x \in \mathsf{X}$, the set of admissible actions is $\mathsf{A}(x) = [\kappa_{\min}, x]$ which is not consistent with our assumptions. However, we can (equivalently) reformulate the above problem so that the admissible actions $\mathsf{A}(x)$ will become A for all $x \in \mathsf{X}$. In this case, instead of dynamics in equation (4.42) we have

$$X_{t+1} = \theta_1 \min(A_t, X_t) \exp\{-\theta_2 \min(A_t, X_t) + V_t\}, \; t = 0, 1, 2, \ldots$$

and $\mathsf{A}(x) = [\kappa_{\min}, \kappa_{\max}]$ for all $x \in \mathsf{X}$. The one-stage reward function is $u(x - a)1_{\{x \geq a\}}$.

Since X is already compact, it is sufficient to discretize $[\kappa_{\min}, \kappa_{\max}]$. The noise process $\{V_t\}$ is a sequence of independent and identically distributed (i.i.d.) random variables which have common density g supported on $[0, \lambda]$. Therefore, the transition probability $p(\,\cdot\,|x, a)$ is given by

$$\begin{aligned} p\big(D|x, a\big) &= \Pr\left\{X_{t+1} \in D \middle| X_t = x, A_t = a\right\} \\ &= \Pr\left\{\theta_1 \min(a, x) \exp\{-\theta_2 \min(a, x) + v\} \in D\right\} \\ &= \int_D g\bigg(\log(v) - \log(\theta_1 \min(a, x)) + \theta_2 \min(a, x)\bigg)\frac{1}{v} m(dv), \end{aligned}$$

for all $D \in \mathscr{B}(\mathbb{R})$. To make the model consistent, we must have $\theta_1 y \exp\{-\theta_2 y + v\} \in [\kappa_{\min}, \kappa_{\max}]$ for all $(y, v) \in [\kappa_{\min}, \kappa_{\max}] \times [0, \lambda]$.

We assume that (i) $g > \epsilon$ for some $\epsilon \in \mathbb{R}_+$ on $[0, \lambda]$, (ii) g is continuous on $[0, \lambda]$, and (iii) the utility function u is continuous. Define $h(v, x, a) := g\big(\log(v) - \log(\theta_1 \min(a, x)) + \theta_2 \min(a, x)\big)\frac{1}{v}$, and for each $(x, a) \in \mathsf{X} \times \mathsf{A}$, let $S_{x,a}$ denote the support of $h(\,\cdot\,, x, a)$. Then Assumption 4.2 holds for this model with $\phi(x, a) = \inf_{v \in S_a} h(v, x, a)$ (provided that it is measurable), $\lambda = m_\kappa$ (the Lebesgue measure restricted on $[\kappa_{\min}, \kappa_{\max}]$), and for some $\alpha \in (0, 1)$.

For the numerical results, we use the following values of the parameters:

$$\theta_1 = 1.1, \; \theta_2 = 0.1, \; \kappa_{\max} = 7, \; \kappa_{\min} = 0.005, \; \lambda = 0.5.$$

We assume that the noise process is uniformly distributed over $[0, 0.5]$. Hence, $g \equiv 1$ on $[0, 0.5]$ and otherwise zero. The utility function u is taken to be the shifted isoelastic utility function [40, Section 4.1]

$$u(z) = 3\big((z + 0.5)^{1/3} - (0.5)^{1/3}\big).$$

We selected 25 different values for the number n of grid points to discretize the state space: $n = 10, 20, 30, \ldots, 250$. The grid points are chosen uniformly over the interval $[\kappa_{\min}, \kappa_{\max}]$. We also uniformly discretize the action space A by using the following number of grid points: $5n = 50, 100, 150, \ldots, 1250$.

We use the relative value iteration algorithm [14, Chapter 4.3.1] to compute the value functions of the finite models. For each n, the finite-state models are constructed as in Section 4.2 by replacing Z with $[\kappa_{\min}, \kappa_{\max}]$ and by setting $\nu_n(\,\cdot\,) = m_\kappa(\,\cdot\,)$.

Figure 4.2 shows the graph of the value functions of the finite models corresponding to the different values of n (number of grid points), when the initial state is $x = 2$. It can be seen that the value functions converge (to the value function of the original model).

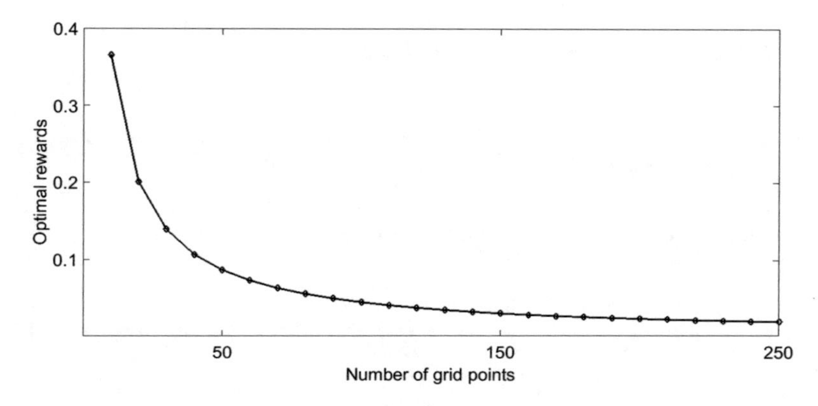

Fig. 4.2 Optimal rewards of the finite models when the initial state is $x = 2$

4.7 Proofs

Proof of Lemma 4.15

We will prove the lemma by induction. Note that if one views the stochastic kernel $p(\,\cdot\,|z,a)$ as a mapping from $\mathsf{Z} \times \mathsf{A}$ to $\mathscr{P}(\mathsf{Z})$, then Assumption 4.2 implies that this mapping is continuous, and therefore uniformly continuous, when $\mathscr{P}(\mathsf{Z})$ is equipped with the metric induced by the total variation distance.

For $t = 1$ the claim holds by the following argument:

$$
\begin{aligned}
&\sup_{(y,f)\in\mathsf{Z}\times\mathbb{F}} \big\| p(\,\cdot\,|y, f(y)) - q_n(\,\cdot\,|y, f(y)) \big\|_{TV} \\
&\qquad := 2 \sup_{(y,f)\in\mathsf{Z}\times\mathbb{F}} \sup_{D\in\mathscr{B}(\mathsf{Z})} \big| p(D|y, f(y)) - q_n(D|y, f(y)) \big| \\
&\qquad \le 2 \sup_{(y,f)\in\mathsf{Z}\times\mathbb{F}} \sup_{D\in\mathscr{B}(\mathsf{Z})} \int \big| p(D|y, f(y)) - p(D|z, f(y)) \big| \, \nu_{n,i_n(y)}(dz) \\
&\qquad \le \sup_{(y,f)\in\mathsf{Z}\times\mathbb{F}} \int \big\| p(\,\cdot\,|y, f(y)) - p(\,\cdot\,|z, f(y)) \big\|_{TV} \nu_{n,i_n(y)}(dz) \\
&\qquad \le \sup_{y\in\mathsf{Z}} \sup_{(z,a)\in\mathscr{S}_{n,i_n(y)}\times\mathsf{A}} \big\| p(\,\cdot\,|y, a) - p(\,\cdot\,|z, a) \big\|_{TV}.
\end{aligned}
$$

As the mapping $p(\,\cdot\,|z, a) : \mathsf{Z} \times \mathsf{A} \to \mathscr{P}(\mathsf{Z})$ is uniformly continuous with respect to the total variation distance and $\max_i \operatorname{diam}(\mathscr{S}_{n,i}) \to 0$ as $n \to \infty$, the result follows. Assume the claim is true for $t \geq 1$. Then we have

$$\sup_{(y,f)\in\mathsf{Z}\times\mathbb{F}} \big\| p^{t+1}(\,\cdot\,|y, f(y)) - q_n^{t+1}(\,\cdot\,|y, f(y)) \big\|_{TV}$$

$$:= \sup_{(y,f)\in\mathsf{Z}\times\mathbb{F}} \sup_{\|g\|\leq 1} \left| \int_{\mathsf{Z}} g(x) p^{t+1}(dx|y, f(y)) - \int_{\mathsf{Z}} g(x) q_n^{t+1}(dx|y, f(y)) \right|$$

$$\leq \sup_{(y,f)\in\mathsf{Z}\times\mathbb{F}} \Bigg(\sup_{\|g\|\leq 1} \bigg| \int_{\mathsf{Z}} \int_{\mathsf{Z}} g(x) p(dx|z, f(z)) p^t(dz|y, f(y))$$

$$- \int_{\mathsf{Z}} \int_{\mathsf{Z}} g(x) p(dx|z, f(z)) q_n^t(dz|y, f(y)) \bigg|$$

$$+ \sup_{\|g\|\leq 1} \bigg| \int_{\mathsf{Z}} \int_{\mathsf{Z}} g(x) p(dx|z, f(z)) q_n^t(dz|y, f(y))$$

$$- \int_{\mathsf{Z}} \int_{\mathsf{Z}} g(x) q_n(dx|z, f(z)) q_n^t(dz|y, f(y)) \bigg| \Bigg)$$

$$\leq \sup_{(y,f)\in\mathsf{Z}\times\mathbb{F}} \big\| p^t(\,\cdot\,|y, f(y)) - q_n^t(\,\cdot\,|y, f(y)) \big\|_{TV}$$

$$+ \sup_{(z,f)\in\mathsf{Z}\times\mathbb{F}} \big\| p(\,\cdot\,|z, f(z)) - q_n(\,\cdot\,|z, f(z)) \big\|_{TV}, \tag{4.43}$$

where the last inequality follows from the following property of the total variation distance: for any $h \in \mathscr{B}(\mathsf{Z})$ and $\mu, \nu \in \mathscr{P}(\mathsf{Z})$ we have $\big| \int_{\mathsf{Z}} h(z)\mu(dz) - \int_{\mathsf{Z}} h(z)\nu(dz) \big| \leq \|h\| \|\mu - \nu\|_{TV}$. By the first step of the proof and the induction hypothesis, the last term converges to zero as $n \to \infty$. This completes the proof.

Proof Lemma 4.22

It is straightforward to prove (4.9) by using the definitions of b_n and w_n, and equation (3.7). To prove (4.10), we have to consider two cases: $x \in K_n$ and $x \in K_n^c$. For the first case, $q_n(\,\cdot\,|x, a) = p(\,\cdot\,|x, a)$, and therefore, we have

$$\sup_{a\in\mathsf{A}} \int_{\mathsf{X}} w_n(y) p(dy|x, a) = \sup_{a\in\mathsf{A}} \bigg\{ \int_{\mathsf{X}} w(y) p(dy|x, a) + \int_{K_n^c} \big(\gamma_n - w(y)\big)\ p(dy|x, a) \bigg\}$$

$$\leq \sup_{a\in\mathsf{A}} \int_{\mathsf{X}} w(y) p(dy|x, a) + \gamma \ \text{ (by (4.5))}$$

$$\leq \alpha w(x) + \gamma = \alpha w_n(x) + \gamma \ \text{ (as } w_n = w \text{ on } K_n).$$

For $x \in K_n^c$, we have

$$\begin{aligned}\sup_{a\in\mathsf{A}} \int_{\mathsf{X}} w_n(y) q_n(dy|x,a) &= \sup_{a\in\mathsf{A}} \int_{K_n^c} \biggl(\int_{\mathsf{X}} w_n(y) p(dy|z,a) \biggr) \vartheta_n(dz) \\ &\leq \int_{K_n^c} \biggl(\sup_{a\in\mathsf{A}} \int_{\mathsf{X}} w_n(y) p(dy|z,a) \biggr) \vartheta_n(dz) \\ &\leq \int_{K_n^c} \bigl(\alpha w(z) + \gamma \bigr) \, \vartheta_n(dz) \qquad (4.44) \\ &= \alpha w_n(x) + \gamma,\end{aligned}$$

where (4.44) can be proved following the same argument as for the case $x \in K_n$. This completes the proof.

Proof of Lemma 4.30
The proof of the first inequality follows from Assumption 4.5 and definitions of b_n and w_n. To prove the remaining two inequalities, we have to consider the cases $x \in K_n$ and $x \in K_n^c$ separately.

Let $x \in K_n$, and therefore, $q_n(\,\cdot\,|x,a) = p(\,\cdot\,|x,a)$. The second inequality holds since

$$\begin{aligned}\int_{\mathsf{X}} w_n(y) p(dy|x,a) &= \int_{\mathsf{X}} w(y) p(dy|x,a) + \int_{K_n^c} \bigl(\gamma_n - w(y) \bigr) \, p(dy|x,a) \\ &\leq \int_{\mathsf{X}} w(y) p(dy|x,a) + \tau_n \\ &\leq \alpha w(x) + \lambda(w) \phi(x,a) + \tau_n \\ &\leq \alpha w_n(x) + \lambda(w_n) \phi_n(x,a) + \varsigma_n \phi_n(x,a) + \tau_n \text{ (as } w_n = w \text{ and } \phi_n = \phi \text{ on } K_n) \\ &\leq \alpha_n w_n(x) + \lambda(w_n) \phi_n(x,a), \ \ \text{(as } \phi_n \leq 1 \text{ and } w_n \geq 1).\end{aligned}$$

For the last inequality, for all $D \in \mathscr{B}(\mathsf{X})$, we have

$$q_n(D|x,a) = p(D|x,a) \geq \lambda(D) \phi(x,a) = \lambda(D) \phi_n(x,a) \ \ (\text{as } \phi_n = \phi \text{ on } K_n).$$

Hence, the inequalities hold for $x \in K_n$.

For $x \in K_n^c$, we have

$$\begin{aligned}\int_{\mathsf{X}} w_n(y) q_n(dy|x,a) &= \int_{K_n^c} \biggl(\int_{\mathsf{X}} w_n(y) p(dy|z,a) \biggr) \vartheta_n(dz) \\ &\leq \int_{K_n^c} \bigl(\alpha w(z) + \lambda(w_n) \phi(x,a) + \varsigma_n \phi(x,a) + \tau_n \bigr) \, \vartheta_n(dz) \\ & \qquad (4.45)\end{aligned}$$

$$= \alpha w_n(x) + \lambda(w_n)\phi_n(x,a) + \varsigma_n \phi_n(x,a) + \tau_n$$
$$\leq \alpha_n w_n(x) + \lambda(w_n)\phi_n(x,a), \;\; \text{(since } \phi_n \leq 1 \text{ and } w_n \geq 1\text{)},$$

where (4.45) can be obtained following the same arguments as for the case $x \in K_n$. The last inequality holds for $x \in K_n^c$ since

$$\begin{aligned} q_n(D|x,a) &= \int_{K_n^c} p(D|z,a)\vartheta_n(dz) \\ &\geq \int_{K_n^c} \lambda(D)\phi(z,a)\vartheta_n(dz) \\ &= \lambda(D)\phi_n(x,a). \end{aligned}$$

This completes the proof.

Proof of Lemma 4.31

We will prove the lemma by induction. Fix any compact set $K \subset \mathsf{X}$. We note that in the inequalities below, we repeatedly use the fact $\phi, \phi_n \leq 1$ without explicitly referring to it. Recall the definition of the compact subsets K_ε of X in Lemma 3.17 and the constant $\gamma_{\max} = \max\{1, \gamma\}$. Note that $\sup_{a \in \mathsf{A}} |g_n(x,a)| \leq M_g w_n(x) \leq M_g \gamma_{\max} w(x)$ for all $x \in \mathsf{X}$.

The claim holds for $t = 1$ by the following argument:

$$\sup_{(y,f) \in K \times \mathbb{F}} \biggl| \int_{\mathsf{X}} g_n(x, f(x)) q_n(dx|y, f(y)) - \int_{\mathsf{X}} g(x, f(x)) p(dx|y, f(y)) \biggr|$$
$$= \sup_{(y,f) \in K \times \mathbb{F}} \biggl| \int_{\mathsf{X}} g_n(x, f(x)) p(dx|y, f(y)) - \int_{\mathsf{X}} g(x, f(x)) p(dx|y, f(y)) \biggr|$$
$$\text{(for } n \text{ sufficiently large)}$$
$$= \sup_{(y,f) \in K \times \mathbb{F}} \biggl| \int_{K_\varepsilon^c} g_n(x, f(x)) p(dx|y, f(y)) - \int_{K_\varepsilon^c} g(x, f(x)) p(dx|y, f(y)) \biggr|$$
$$\text{(for } n \text{ sufficiently large)}$$
$$\leq M_g(1 + \gamma_{\max})\varepsilon,$$

where the last inequality follows from Lemma 3.17. Since ε is arbitrary, the result follows.

Assume the claim is true for $t \geq 1$. Let us define $l_f(z) := \int_{\mathsf{X}} g(x, f(x)) p^t(dx|z, f(z))$ and $l_f^n(z) := \int_{\mathsf{X}} g_n(x, f(x)) q_n^t(dx|z, f(z))$. By recursively applying the inequalities in Assumption 4.5-(c) and in (4.27), we obtain

$$\sup_{f \in \mathbb{F}} |l_f(z)| \leq M_g \alpha^t w(z) + M_g \eta(w) \sum_{j=0}^{t-1} \alpha^j$$

and

$$\sup_{f \in \mathbb{F}} |l_f^n(z)| \leq M_g \alpha_n^t w_n(z) + M_g \eta(w_n) \sum_{j=0}^{t-1} \alpha_n^j$$

$$\leq M_g \alpha_{\max}^t \gamma_{\max} w(z) + M_g \eta(w) \gamma_{\max} \sum_{j=0}^{t-1} \alpha_{\max}^j,$$

where $\alpha_{\max} := \sup_{n \geq n_0} \alpha_n < 1$. Then we have

$$\sup_{(y,f) \in K \times \mathbb{F}} \left| \int_{\mathsf{X}} g_n(x, f(x)) q_n^{t+1}(dx|y, f(y)) - \int_{\mathsf{X}} g(x, f(x)) p^{t+1}(dx|y, f(y)) \right|$$

$$= \sup_{(y,f) \in K \times \mathbb{F}} \left| \int_{\mathsf{X}} l_f^n(z) q_n(dz|y, f(y)) - \int_{\mathsf{X}} l_f(z) p(dz|y, f(y)) \right|$$

$$= \sup_{(y,f) \in K \times \mathbb{F}} \left| \int_{\mathsf{X}} l_f^n(z) p(dz|y, f(y)) - \int_{\mathsf{X}} l_f(z) p(dz|y, f(y)) \right| \text{ (for } n \text{ sufficiently large)}$$

$$\leq \sup_{(y,f) \in K \times \mathbb{F}} \left| \int_{K_\varepsilon^c} l_f^n(z) p(dz|y, f(y)) - \int_{K_\varepsilon^c} l_f(z) p(dz|y, f(y)) \right|$$

$$+ \sup_{(z,f) \in K_\varepsilon \times \mathbb{F}} |l_f^n(z) - l_f(z)| \leq L\varepsilon + \sup_{(z,f) \in K_\varepsilon \times \mathbb{F}} |l_f^n(z) - l_f(z)|, \tag{4.46}$$

where L is given by

$$L := M_g \left(\alpha^t + \alpha_{\max}^t \gamma_{\max} + \eta(w) \sum_{j=0}^{t-1} \alpha^j + \eta(w) \gamma_{\max} \sum_{j=0}^{t-1} \alpha_{\max}^j \right)$$

and the last inequality follows from Lemma 3.17. Since the claim holds for t and K_ε, the second term in (4.46) goes to zero as $n \to \infty$. Since ε is arbitrary, the result follows.

Proof of Lemma 4.32

To ease the notation, we define $M(\mathsf{X}_n)$, $M(\mathsf{X})$, and $M_w(\mathsf{X})$ as the subsets of $B(\mathsf{X}_n)$, $B(\mathsf{X})$, and $B_w(\mathsf{X})$, respectively, whose elements have (corresponding) norm less than one. Let $(x_k, a_k) \to (x, a)$ in $\mathsf{X}_n \times \mathsf{A}$. Since the pseudo state Δ_n is isolated and K_n is compact, we have two cases: (i) $x_k = x = \Delta_n$ for all k large enough, or (ii) $x_k \to x$ in K_n.

For the first case we have

$$\|p_n(\,\cdot\,|\Delta_n, a_k) - p_n(\,\cdot\,|\Delta_n, a)\|_{TV}$$

$$= \sup_{g \in M(\mathsf{X}_n)} \left| \int_{\mathsf{X}_n} g(y) p_n(dy|\Delta_n, a_k) - \int_{\mathsf{X}_n} g(y) p_n(dy|\Delta_n, a) \right|$$

$$
\begin{aligned}
&\leq \sup_{g\in M(\mathsf{X})} \left| \int_{\mathsf{X}} g(y) q_n(dy|\Delta_n, a_k) - \int_{\mathsf{X}} g(y) q_n(dy|\Delta_n, a) \right| && (4.47)\\
&= \sup_{g\in M(\mathsf{X})} \left| \int_{K_n^c} \left(\int_{\mathsf{X}} g(y) p(dy|z, a_k) - \int_{\mathsf{X}} g(y) p(dy|z, a) \right) \vartheta_n(dz) \right| \\
&\leq \int_{K_n^c} \sup_{g\in M(\mathsf{X})} \left| \int_{\mathsf{X}} g(y) p(dy|z, a_k) - \int_{\mathsf{X}} g(y) p(dy|z, a) \right| \vartheta_n(dz) \\
&\leq \int_{K_n^c} \sup_{g\in M_w(\mathsf{X})} \left| \int_{\mathsf{X}} g(y) p(dy|z, a_k) - \int_{\mathsf{X}} g(y) p(dy|z, a) \right| \vartheta_n(dz) \\
&= \int_{K_n^c} \| p(\,\cdot\,|z, a_k) - p(\,\cdot\,|z, a) \|_w \vartheta_n(dz), && (4.48)
\end{aligned}
$$

where (4.47) follows since if for any $g \in M(\mathsf{X}_n)$ we define $\bar{g} = g$ on K_n and $\bar{g} = g(\Delta_n)$ on K_n^c, then we have $\bar{g} \in M(\mathsf{X})$ and $\int_{\mathsf{X}_n} g(y) p_n(dy|x, a) = \int_{\mathsf{X}} \bar{g}(y) q_n(dy|x, a)$ for all $(x, a) \in \mathsf{X}_n \times \mathsf{A}$. Note that we have

$$
\begin{aligned}
&\sup_{g\in M_w(\mathsf{X})} \left| \int_{\mathsf{X}} g(y) p(dy|z, a_k) - \int_{\mathsf{X}} g(y) p(dy|z, a) \right| \\
&\leq \int_{\mathsf{X}} w(y) p(dy|z, a_k) + \int_{\mathsf{X}} w(y) p(dy|z, a) \\
&\leq 2\big(\alpha + \lambda(w)\big) w(z) \quad \text{(by Assumption 4.5-(c), } \phi \leq 1, \text{ and } w \geq 1).
\end{aligned}
$$

Since w (restricted to K_n^c) is ϑ_n-integrable, by the dominated convergence theorem (4.48) goes to zero as $k \to \infty$.

For the second case we have

$$
\begin{aligned}
&\| p_n(\,\cdot \mid x_k, a_k) - p_n(\,\cdot\,|x, a) \|_{TV} \\
&\quad = \sup_{g\in M(\mathsf{X}_n)} \left| \int_{\mathsf{X}_n} g(y) p_n(dy|x_k, a_k) - \int_{\mathsf{X}_n} g(y) p_n(dy|x, a) \right| \\
&\quad \leq \sup_{g\in M(\mathsf{X})} \left| \int_{\mathsf{X}} g(y) q_n(dy|x_k, a_k) - \int_{\mathsf{X}} g(y) q_n(dy|x, a) \right| \\
&\quad = \sup_{g\in M(\mathsf{X})} \left| \int_{\mathsf{X}} g(y) p(dy|x_k, a_k) - \int_{\mathsf{X}} g(y) p(dy|x, a) \right| \quad \text{(since } x_k, x \in K_n) \\
&\quad \leq \sup_{g\in M_w(\mathsf{X})} \left| \int_{\mathsf{X}} g(y) p(dy|x_k, a_k) - \int_{\mathsf{X}} g(y) p(dy|x, a) \right| \\
&\quad = \| p(\,\cdot\,|x_k, a_k) - p(\,\cdot\,|x, a) \|_w.
\end{aligned}
$$

By Assumption 4.5-(b) the last term goes to zero as $k \to \infty$.

Proof of Lemma 4.41
Similar to the proof of Lemma 4.15, we use induction. For $t = 1$, recalling the proof of Lemma 4.15, the claim holds by the following argument:

$$\sup_{(y,f)\in \mathsf{Z}\times\mathbb{F}} \|p(\cdot\,|y, f(y)) - q_n(\cdot\,|y, f(y))\|_{TV} \leq \sup_{y\in\mathsf{Z}} \sup_{(x,a)\in\mathscr{S}_{n,i_n(y)}\times\mathsf{A}} \|p(\cdot\,|y, a) - p(\cdot\,|x, a)\|_{TV}$$
$$\leq \omega_p(d_n).$$

Now, assume the claim is true for $t \geq 1$. Again recalling the proof of Lemma 4.15, we have

$$\sup_{(y,f)\in \mathsf{Z}\times\mathbb{F}} \|p^{t+1}(\cdot\,|y, f(y)) - q_n^{t+1}(\cdot\,|y, f(y))\|_{TV}$$
$$\leq \sup_{(y,f)\in \mathsf{Z}\times\mathbb{F}} \left\|p^{t}(\cdot\,|y, f(y)) - q_n^{t}(\cdot\,|y, f(y))\right\|_{TV}$$
$$+ \sup_{(z,f)\in \mathsf{Z}\times\mathbb{F}} \left\|p(\cdot\,|z, f(z)) - q_n(\cdot\,|z, f(z))\right\|_{TV}$$
$$\leq t\omega_p(d_n) + \omega_p(d_n) = (t+1)\omega_p(d_n).$$

This completes the proof.

4.8 Concluding Remarks

In this chapter, the approximation of a discrete-time MDP by finite-state MDPs was considered for discounted and average costs for both compact and non-compact state spaces. Under conditions usually imposed in studying Markov decision processes, it was shown that if one uses a sufficiently large number of grid points to discretize the state space, then the resulting finite-state MDP yields a near optimal policy. Under the Lipschitz continuity of the transition probability and the one-stage cost function, explicit bounds were derived on the performance loss due to discretization in terms of the number of grid points for the compact state case. These results were then illustrated numerically by considering two different MDP models. The results of this chapter are based on the papers [115, 116, 119].

We refer the reader to [41, 42] for a further reading on finite model approximation problem for MDPs with general state and action spaces. In [42], the authors considered the approximation problem for average cost Markov decision processes. They adopt a simulation based approximation technique leading to probabilistic guarantees on the approximation error. In [41], the authors consider the approximation problem for constrained Markov decision processes with discounted cost, and obtain approximation results analogous with [42] using a linear programming formulation.

Chapter 5
Approximations for Partially Observed Markov Decision Processes

This chapter studies the finite-model approximation of discrete-time partially observed Markov decision process. We will find that by performing the standard reduction method, where one transforms a partially observed model to a belief-based fully observed model, we can apply and properly generalize the results in the preceding chapters to obtain approximation results. The versatility of approximation results under weak continuity conditions become particularly evident while investigating the applicability of these results to the partially observed case. We also provide systematic procedures for the quantization of the set of probability measures on the state space of POMDPs which is the state space of belief-MDPs.

5.1 Introduction

Recall from Section 2.2 that a discrete-time partially observed Markov decision process (POMDP) is specified by

$$\big(\mathsf{X}, \mathsf{A}, \mathsf{Y}, p, r, c\big),$$

where $\big(\mathsf{X}, \mathsf{A}, \mathsf{Y}, p, c\big)$ are the components of a usual MDP and the stochastic kernel $r : \mathsf{X} \to \mathscr{P}(\mathsf{Y})$ denotes the transition probability of the current observation given the current state (i.e., *observation channel*). In a POMDP, the decision maker can only use its observation-action history when constructing its control policy. Namely, a *policy* $\pi = \{\pi_t\}$ is a sequence of stochastic kernels on A given $\mathsf{H}_t = (\mathsf{Y} \times \mathsf{A})^t \times \mathsf{Y}$, where H_t contains observation-action history up to time t.

N. Saldi et al., *Finite Approximations in Discrete-Time Stochastic Control*, Systems & Control: Foundations & Applications,
https://doi.org/10.1007/978-3-319-79033-6_5

As indicated in Section 2.2, any POMDP can be reduced to a (completely observable) MDP [108, 151], whose states are the posterior state distributions or "beliefs" of the observer; that is, the state at time t is

$$Z_t := \mathsf{Pr}\{X_t \in \cdot \,|Y_0, \ldots, Y_t, A_0, \ldots, A_{t-1}\} \in \mathscr{P}(\mathsf{X}).$$

This equivalent MDP is called the belief-MDP.

In this chapter, we apply the approximation results obtained in Sections 3.4.1 and 4.3.1 for the discounted cost criterion to partially observed Markov decision processes (POMDPs) via the fully observed reduction technique. To this end, let $v : \mathsf{X} \to [0, \infty)$ be a continuous "moment function" having the property that there exists an increasing sequence of compact subsets $\{K_n\}_{n\geq 1}$ of X such that

$$\lim_{n\to\infty} \inf_{x\in\mathsf{X}\setminus K_n} v(x) = \infty.$$

The following assumptions will be imposed on the components of the POMDP.

Assumption 5.1

(a) The one-stage cost function c is continuous and bounded.
(b) The stochastic kernel $p(\,\cdot\,|x, a)$ is weakly continuous in $(x, a) \in \mathsf{X} \times \mathsf{A}$.
(c) The observation channel $r(\,\cdot\,|x)$ is continuous in total variation, i.e., if $x_k \to x$, then $r(\,\cdot\,|x_k) \to r(\,\cdot\,|x)$ in total variation.
(d) A is compact.
(e) There exists a constant $\lambda \geq 0$ such that

$$\sup_{a\in\mathsf{A}} \int_{\mathsf{X}} v(y)p(dy|x, a) \leq \lambda v(x).$$

(f) The initial probability measure μ satisfies

$$\int_{\mathsf{X}} v(x)\mu(dx) < \infty.$$

We let

$$\mathscr{P}_v(\mathsf{X}) := \biggl\{\mu \in \mathscr{P}(\mathsf{X}) : \int_{\mathsf{X}} v(x)\mu(dx) < \infty\biggr\}.$$

Note that since the probability law of X_t is in $\mathscr{P}_v(\mathsf{X})$, by Assumption 5.1-(e),(f), under any policy we have $\mathsf{Pr}\{X_t \in \cdot\,|Y_0, \ldots, Y_t, A_0, \ldots, A_{t-1}\} \in \mathscr{P}_v(\mathsf{X})$ almost everywhere. Therefore, the belief-MDP has state space $\mathsf{Z} = \mathscr{P}_v(\mathsf{X})$ instead of $\mathscr{P}(\mathsf{X})$, where Z is equipped with the Borel σ-algebra generated by the topology of weak convergence. The transition probability η of the belief-MDP can be constructed as follows (see also [66]). Let z denote the generic state variable for the belief-MDP. First consider the transition probability on $\mathsf{X} \times \mathsf{Y}$ given $\mathsf{Z} \times \mathsf{A}$

$$R(x \in A, y \in B|z, a) := \int_{\mathsf{X}} \kappa(A, B|x', a) z(dx'),$$

where $\kappa(dx, dy|x', a) := r(dy|x) \otimes p(dx|x', a)$. Let us disintegrate R as

$$R(dx, dy|z, a) = H(dy|z, a) \otimes F(dx|z, a, y).$$

Then, we define the mapping $F : \mathsf{Z} \times \mathsf{A} \times \mathsf{Y} \to \mathsf{Z}$ as

$$F(z, a, y) = F(\,\cdot\,|z, a, y). \tag{5.1}$$

In the literature, (5.1) is called the "nonlinear filtering equation". Note that, for each $t \geq 0$, we indeed have

$$F(z, a, y)(\,\cdot\,) = \Pr\{X_{t+1} \in \,\cdot\,|Z_t = z, A_t = a, Y_{t+1} = y\}$$

and

$$H(\,\cdot\,|z, a) = \Pr\{Y_{t+1} \in \,\cdot\,|Z_t = z, A_t = a\}.$$

Then, η can be written as

$$\eta(\,\cdot\,|z, a) = \int_{\mathsf{Y}} \delta_{F(z,a,y)}(\,\cdot\,)\, H(dy|z, a),$$

where z denotes the posterior distribution of the state X_t given the past observations. The initial point for the belief-MDP is μ; that is, $Z_0 \sim \delta_{\mu}$. Finally, the one-stage cost function $\tilde{c}$ of the belief-MDP is given by

$$\tilde{c}(z, a) := \int_{\mathsf{X}} c(x, a) z(dx). \tag{5.2}$$

Hence, the belief-MDP is a Markov decision process with the components

$$(\mathsf{Z}, \mathsf{A}, \eta, \tilde{c}).$$

For the belief-MDP define the history spaces $\tilde{\mathsf{H}}_0 = \mathsf{Z}$ and $\tilde{\mathsf{H}}_t = (\mathsf{Z} \times \mathsf{A})^t \times \mathsf{Z}$, $t = 1, 2, \ldots$ and let $\tilde{\Pi}$ denote the set of all policies for the belief-MDP, where the policies are defined in the usual manner. Let $\tilde{W}(\tilde{\pi}, \xi)$ denote the cost function of policy $\tilde{\pi} \in \tilde{\Pi}$ for initial distribution ξ of the belief-MDP, where $\tilde{W} \in \{J, V\}$.

Notice that any history vector $\tilde{h}_t = (z_0, \ldots, z_t, a_0, \ldots, a_{t-1})$ of the belief-MDP is a function of the history vector $h_t = (y_0, \ldots, y_t, a_0, \ldots, a_{t-1})$ of the POMDP. Let us write this relation as $i(h_t) = \tilde{h}_t$. Hence, for a policy $\tilde{\pi} = \{\tilde{\pi}_t\} \in \tilde{\Pi}$, we can define a policy $\pi^{\tilde{\pi}} = \{\pi_t^{\tilde{\pi}}\} \in \Pi$ as

$$\pi_t^{\tilde{\pi}}(\cdot|h_t) := \tilde{\pi}_t(\cdot|i(h_t)). \tag{5.3}$$

Let us write this as a mapping from $\tilde{\Pi}$ to Π: $\tilde{\Pi} \ni \tilde{\pi} \mapsto i(\tilde{\pi}) = \pi^{\tilde{\pi}} \in \Pi$. It is straightforward to show that the cost functions $\tilde{W}(\tilde{\pi}, \xi)$ and $W(\pi^{\tilde{\pi}}, \mu)$ are the same, where $\xi = \delta_\mu$. One can also prove that (see [108, 151])

$$\inf_{\tilde{\pi} \in \tilde{\Pi}} \tilde{W}(\tilde{\pi}, \xi) = \inf_{\pi \in \Pi} W(\pi, \mu) \tag{5.4}$$

and furthermore that if $\tilde{\pi}$ is an optimal policy for the belief-MDP, then $\pi^{\tilde{\pi}}$ is optimal for the POMDP. Hence, the POMDP and the corresponding belief-MDP are equivalent in the sense of cost minimization. Therefore, approximation results developed for MDPs in the previous chapters can be applied to the belief-MDP and so, to the POMDP. However, for the belief-MDP, the regularity conditions imposed on the transition probability in Sections 3.4.2 and 4.3.2 for the average cost criterion do not hold in general. Indeed, as will be seen in the next section, even the strong continuity condition is prohibitive for the transition probability of the belief-MDP. Therefore, in this chapter, we will restrict our attention to the discounted cost.

5.2 Continuity Properties of the Belief-MDP

In this section, we first discuss continuity properties that are satisfied by or prohibitive for the transition probability η of the belief-MDP. Then, we derive conditions satisfied by the components of the belief-MDP.

5.2.1 *Strong and Weak Continuity Properties*

Recall that the stochastic kernel η is said to be *weakly* continuous if $\eta(\cdot|z_k, a_k) \to \eta(\cdot|z, a)$ weakly, whenever $(z_k, a_k) \to (z, a)$. The kernel is said to be *strongly* continuous if, for any $z \in \mathsf{Z}$, $\eta(\cdot|z, a_k) \to \eta(\cdot|z, a)$ setwise, whenever $a_k \to a$. For the fully observed reduction of a partially observed MDP (POMDP), strong continuity of the transition probability is in general a prohibitive condition. This is illustrated through the following simple example.

Example 5.1 Consider the system dynamics

$$X_{t+1} = X_t + A_t,$$
$$Y_t = X_t + V_t,$$

where $X_t \in \mathsf{X}$, $Y_t \in \mathsf{Y}$, $A_t \in \mathsf{A}$, and where X, Y, and A are the state, observation, and action spaces, respectively, all of which we take to be $\mathbb{R}_+$ (the nonnegative real line) and the "noise" process $\{V_t\}$ is a sequence of i.i.d. random variables uniformly distributed on $[0, 1]$. It is easy to see that the transition probability $p(\cdot|x, a) := \Pr\{X_{t+1} \in \cdot|X_t = x, A_t = a\}$ is weakly continuous with respect to state-action variables (x, a) and the observation channel $r(\cdot|x) := \Pr\{Y_t \in \cdot|X_t = x\}$ is continuous in total variation with respect to state variable x for this POMDP. Hence, by [50, Theorem 3.6] the transition probability η of the belief-MDP is weakly continuous in the state-action variables. However, η is not setwise continuous with respect to the action variable as shown below.

Recall that η is given by

$$\eta(\cdot|z, a) = \int_{\mathsf{Y}} 1_{\{F(z,a,y)\in\cdot\}} H(dy|z, a),$$

where $F(z, a, y)(\cdot) = \Pr\{X_{t+1} \in \cdot|Z_t = z, A_t = a, Y_{t+1} = y\}$, $H(\cdot|z, a) = \Pr\{Y_{t+1} \in \cdot|Z_t = z, A_t = a\}$, and $Z_t \in \mathsf{Z} = \mathscr{P}(\mathsf{X})$ is the posterior distribution of the state X_t given the past observations, i.e.,

$$Z_t(\cdot) = \Pr\{X_t \in \cdot|Y_0, \ldots, Y_t, A_0, \ldots, A_{t-1}\}.$$

Let us set $z = \delta_0$ (point mass at 0), $\{a_k\} = \{\frac{1}{k}\}$, and $a = 0$. Then $a_k \to a$, but as we next show, $\eta(\cdot|z, a_k)$ does not converge to $\eta(\cdot|z, a)$ setwise.

Observe that for all k and $y \in \mathsf{Y}$, we have $F(z, a_k, y) = \delta_{\frac{1}{k}}$ and $F(z, a, y) = \delta_0$. Define the open set O with respect to the weak topology in Z as

$$O := \left\{z \in \mathsf{Z} : \left| \int_{\mathsf{X}} g(x)\delta_1(dx) - \int_{\mathsf{X}} g(x)z(dx) \right| < 1 \right\},$$

where $g(x) = |1 - x|$ if $x \in [-1, 1]$ and $g(x) = 0$ otherwise. Observe that we have $F(z, a_k, y) \in O$ for all k and y, but $F(z, a, y) \notin O$ for all y. Hence,

$$\eta(O|z, a_k) := \int_{\mathsf{Y}} 1_{\{F(z,a_k,y)\in O\}} H(dy|z, a_k) = 1,$$

but

$$\eta(O|z, a) := \int_{\mathsf{Y}} 1_{\{F(z,a,y)\in O\}} H(dy|z, a) = 0,$$

implying that $\eta(\cdot\,|z, a_k)$ does not converge to $\eta(\cdot\,|z, a)$ setwise. Hence, η does not satisfy the strong continuity assumption.

The following theorem is a consequence of [50, Theorem 3.6, Example 4.1] and the preceding example.

Theorem 5.2

(i) Under Assumption 5.1-(b),(c), the stochastic kernel η for belief-MDP is weakly continuous in (z, a).

(ii) If we relax the continuity in total variation of the observation channel to setwise or weak continuity, then η may not be weakly continuous even if the transition probability p of POMDP is continuous in total variation.

(iii) Finally, η may not be setwise continuous in action variable a even if the observation channel is continuous in total variation.

Part (i) of Theorem 5.2 implies that the transition probability η of the belief-MDP is weakly continuous under Assumption 5.1. However, note that continuity of the observation channel in total variation in Assumption 5.1 cannot be relaxed to weak or setwise continuity. On the other hand, the continuity of the observation channel in total variation is not enough for the setwise continuity of η. Hence, the results in Section 3.3 cannot be applied to the POMDP we consider even if we put a fairly strong condition on the observation channel.

The above suggest that our earlier results in Sections 3.4.1 and 4.3.1, which only require weak continuity conditions on the transition kernel of a given MDP, are particularly suitable in developing approximation methods for POMDPs (through their MDP reduction), in quantizing both the action spaces and the state spaces.

Remark 5.3 We refer the reader to [50, Theorem 3.2(c)] for more general conditions implying weak continuity of the transition probability η. We also note that, in the uncontrolled setting, [18] and [27] have established similar weak continuity conditions (i.e., the weak-Feller property) of the nonlinear filter process (i.e., the belief process) in continuous time and discrete time, respectively.

Example 5.4 In this example we consider the following partially observed model

$$\begin{aligned} X_{t+1} &= F(X_t, A_t, V_t), \\ Y_t &= H(X_t, W_t),\ t = 0, 1, 2, \ldots \end{aligned} \tag{5.5}$$

where $\mathsf{X} = \mathbb{R}^n$, $\mathsf{A} \subset \mathbb{R}^m$, and $\mathsf{Y} \subseteq \mathbb{R}^d$ for some $n, m, d \geq 1$. The noise processes $\{V_t\}$ and $\{W_t\}$ are sequences of independent and identically distributed (i.i.d.) random vectors taking values in $\mathsf{V} = \mathbb{R}^p$ and $\mathsf{W} = \mathbb{R}^l$, respectively, for some $p, l \geq 1$, and they are also independent of each other. In this system, the continuity of F in (x, a) is sufficient to imply the weak continuity of the transition probability p, and no assumptions are needed on the noise process (not even the existence of a density is required). On the other hand, the continuity of the observation channel r in total variation holds, if for any $x \in \mathsf{X}$, the probability measure $r(\cdot\,|x)$

has a density $g(y, x)$, which is continuous in x, with respect to some reference probability measure m on Y. This follows from Scheffé's theorem (see, e.g., [19, Theorem 16.2]). For instance, this density condition holds for the following type of models:

(i) In the first model, we have $\mathsf{Y} = \mathsf{W} = \mathbb{R}^d$, $H(x, w) = H(x) + w$, H is continuous, and w has a continuous density g_w with respect to Lebesgue measure.
(ii) In the second case, Y is countable and $r(y|x)$ is continuous in x for all $y \in \mathsf{Y}$.

Therefore, the transition probability η of the belief space MDP, corresponding to the model in (5.5), is weakly continuous.

Next, we derive conditions satisfied by the components of the belief-MDP under Assumption 5.1. Note first that $\mathsf{Z} = \bigcup_{m \geq 1} F_m$ where

$$F_m := \left\{ \mu \in \mathscr{P}_v(\mathsf{X}) : \int_{\mathsf{X}} v(x) \mu(dx) \leq m \right\}.$$

Since v is a moment function, each F_m is tight [68, Proposition E.8]. Moreover, each F_m is also closed since v is continuous. Therefore, each F_m is compact with respect to the weak topology. This implies that Z is a σ-compact Borel space. Note that by [16, Proposition 7.30], the one-stage cost function $\tilde{c}$ of the belief-MDP, which is defined in (5.2), is in $C_b(\mathsf{Z} \times \mathsf{A})$ under Assumption 5.1-(a). Therefore, the belief-MDP satisfies the following conditions under Assumption 5.1, which we formally state as a separate assumption.

Assumption 5.2

(i) The one-stage cost function $\tilde{c}$ is bounded and continuous.
(ii) The stochastic kernel η is weakly continuous.
(iii) A is compact and Z is σ-compact.

5.2.2 *Finite-Action Model*

In this section, we consider finite-action approximation of the belief-MDP and so, the POMDP. For these equivalent models, we obtain an approximate finite-action model exactly as in Section 3.2.1. Indeed, let d_{A} denote the metric on A. Since A is assumed compact and thus totally bounded, there exists a sequence of finite sets $\Lambda_n = \{a_{n,1}, \ldots, a_{n,k_n}\} \subset \mathsf{A}$ such that for each n,

$$\min_{i \in \{1,\ldots,k_n\}} d_{\mathsf{A}}(a, a_{n,i}) < 1/n \text{ for all } a \in \mathsf{A}.$$

In other words, Λ_n is a $1/n$-net in A. The sequence $\{\Lambda_n\}_{n\geq1}$ is used by the finite-action model to approximate the belief-MDP and the POMDP. Given the belief-MDP reduction, an implication of Theorem 3.16 is the following.

Theorem 5.5 *Suppose Assumption 5.1 (and thus Assumption 5.2) holds for the POMDP. Then we have*

$$\lim_{n\to\infty} |\tilde{J}_n^*(z) - \tilde{J}^*(z)| = 0 \ \textit{for all } z \in \mathsf{Z},$$

where $\tilde{J}_n^*$ *is the discounted cost value function of the belief-MDP*$_n$ *with the components* $\big(\mathsf{Z}, \Lambda_n, \eta, \tilde{c}\big)$ *and* $\tilde{J}^*$ *is the discounted cost value function of the belief-MDP with components* $\big(\mathsf{Z}, \mathsf{A}, \eta, \tilde{c}\big)$.

The significance of Theorem 5.5 is reinforced by the following observation. If we let $\Pi(\Lambda_n)$ to be the set of deterministic policies of the POMDP taking values in Λ_n, then the theorem implies that for any given $\varepsilon > 0$ there exists $n \geq 1$ and $\pi^* \in \Pi(\Lambda_n)$ such that

$$J(\pi_\varepsilon^*, \mu) < \min_{\pi\in\Pi} J(\pi, \mu) + \varepsilon,$$

where $\pi_\varepsilon^* = \pi^{\tilde{f}_n^*}$ (see (5.3)) and $\tilde{f}_n^*$ is the optimal deterministic stationary policy for the belief-MDP$_n$. This means that even when there is an information transmission constraint from the controller to the plant, one can get ε-close to the value function for any small ε by quantizing the controller's actions and sending the encoded levels.

5.2.3 *Finite-State Model*

The finite-state model for the belief-MDP is obtained exactly as in Sections 4.2 and 4.3, by quantizing the set of probability measures $\mathsf{Z} = \mathscr{P}_v(\mathsf{X})$; that is, for each m, we quantize compact set F_m similar to the quantization of A and represent the rest of the points $\mathsf{Z} \setminus F_m$ by some pseudo-state. For the sake of completeness, we give detailed finite-model construction below.

If Z is compact, in the following, the index m can be fixed to $m = 1$.

We let d_{Z} denote a metric on Z which metrizes the weak topology. For each $m \geq 1$, since F_m is compact and thus totally bounded, there exists a sequence $\big(\{z_{n,i}^{(m)}\}_{i=1}^{k_n^{(m)}}\big)_{n\geq1}$ of finite grids in F_m such that for all $n \geq 1$,

$$\min_{i\in\{1,\ldots,k_n^{(m)}\}} d_{\mathsf{Z}}(z, z_{n,i}^{(m)}) < 1/n \text{ for all } z \in F_m. \tag{5.6}$$

Let $\{\mathscr{S}_{n,i}^{(m)}\}_{i=1}^{k_n^{(m)}}$ be a partition of F_m such that $z_{n,i}^{(m)} \in \mathscr{S}_{n,i}^{(m)}$ and

$$\max_{z \in \mathscr{S}_{n,i}^{(m)}} d_{\mathsf{Z}}(z, z_{n,i}^{(m)}) < 1/n \tag{5.7}$$

for all $i = 1, \ldots, k_n^{(m)}$. Choose any $z_{n,k_n^{(m)}+1}^{(m)} \in \mathsf{Z} \setminus F_m$ which is a so-called pseudo-state and set $\mathscr{S}_{n,k_n^{(m)}+1} = \mathsf{Z} \setminus F_m$. Let $\mathsf{Z}_n^{(m)} := \{z_{n,1}^{(m)}, \ldots, z_{n,k_n}^{(m)}, z_{n,k_n^{(m)}+1}^{(m)}\}$ and define function $Q_n^{(m)} : \mathsf{Z} \to \mathsf{Z}_n^{(m)}$ by

$$Q_n^{(m)}(z) = z_{n,i}^{(m)} \text{ when } z \in \mathscr{S}_{n,i}^{(m)}.$$

Here $Q_n^{(m)}(z)$ maps z to the representative element of the partition it belongs to.

Remark 5.6

(a) Note that given $\{z_{n,i}^{(m)}\}_{i=1}^{k_n^{(m)}} \subset F_m$ that satisfies (5.6), one way to obtain the corresponding partition $\{\mathscr{S}_{n,i}^{(m)}\}_{i=1}^{k_n^{(m)}}$ of F_m satisfying (5.7) as follows. Let us define function $Q_{\text{near}} : F_m \to \{z_{n,1}^{(m)}, \ldots, z_{n,k_n^{(m)}}^{(m)}\}$ as

$$Q_{\text{near}}(z) = \arg\min_{z_{n,i}^{(m)}} d_{\mathsf{Z}}(z, z_{n,i}^{(m)}),$$

where ties are broken so that Q_{near} is measurable. Then, Q_{near} induces a partition $\{\mathscr{S}_{n,i}^{(m)}\}_{i=1}^{k_n^{(m)}}$ of F_m given by

$$\mathscr{S}_{n,i}^{(m)} = \{z \in F_m : Q_n^{(m)}(z) = z_{n,i}^{(m)}\},$$

and which satisfies (5.7). Although one can construct, in theory, the partition using nearest neighbor sense, it is computationally difficult to find these regions when the original state space X is uncountable.

(b) The index n indicates the resolution of the quantizer that is applied to discretize the compact set F_m and index m emphasizes the size of the compact set F_m for which quantization is applied.

Let $\{\nu_n^{(m)}\}$ be a sequence of probability measures on Z satisfying

$$\nu_n^{(m)}(\mathscr{S}_{n,i}^{(m)}) > 0 \text{ for all } i, n, m. \tag{5.8}$$

One possible choice for $\nu_n^{(m)}$ is

$$\nu_n^{(m)}(\,\cdot\,) = \sum_{i=1}^{k_n^{(m)}+1} \delta_{z_{n,i}^{(m)}}(\,\cdot\,).$$

We let $\nu_{n,i}^{(m)}$ be the restriction of $\nu_n^{(m)}$ to $\mathscr{S}_{n,i}^{(m)}$ defined by

$$\nu_{n,i}^{(m)}(\,\cdot\,) := \frac{\nu_n^{(m)}(\,\cdot\,)}{\nu_n^{(m)}(\mathscr{S}_{n,i}^{(m)})}.$$

The measures $\nu_{n,i}^{(m)}$ will be used to define a sequence of finite-state belief-MDPs, denoted as $\mathrm{MDP}_n^{(m)}$, which approximate the belief-MDP. To this end, for each n and m define the one-stage cost function $c_n^{(m)} : \mathsf{Z}_n^{(m)} \times \mathsf{A} \to [0, \infty)$ and the transition probability $p_n^{(m)}$ on $\mathsf{Z}_n^{(m)}$ given $\mathsf{Z}_n^{(m)} \times \mathsf{A}$ by

$$c_n^{(m)}(z_{n,i}^{(m)}, a) := \int_{\mathscr{S}_{n,i}^{(m)}} c(z, a)\nu_{n,i}^{(m)}(dz),$$

$$p_n^{(m)}(\,\cdot\,|z_{n,i}^{(m)}, a) := \int_{\mathscr{S}_{n,i}^{(m)}} Q_n^{(m)} * p(\,\cdot\,|z, a)\nu_{n,i}^{(m)}(dz), \tag{5.9}$$

where $Q_n^{(m)} * p(\,\cdot\,|z, a) \in \mathscr{P}(\mathsf{Z}_n^{(m)})$ is the pushforward of the measure $p(\,\cdot\,|z, a)$ with respect to $Q_n^{(m)}$. For each n and m, we define $\mathrm{MDP}_n^{(m)}$ as a Markov decision process with the following components: $\mathsf{Z}_n^{(m)}$ is the state space, A is the action space, $p_n^{(m)}$ is the transition probability, and $c_n^{(m)}$ is the one-stage cost function.

Given the belief-MDP, an implication of Theorem 4.27 is the following.

Theorem 5.7 *Suppose Assumption 5.1 (and thus Assumption 5.2) holds for the POMDP. Then we have*

$$\lim_{n,m\to\infty} |\tilde{J}(f_n^{(m)}, \mu) - \tilde{J}^*(\mu)| = 0,$$

where $f_n^{(m)}$ is obtained by extending the optimal policy of the $\mathrm{MDP}_n^{(m)}$ to Z. Hence, by the equivalence of POMDPs and belief-MDPs, we also have

$$\lim_{n,m\to\infty} |J(\pi^{f_n^{(m)}}, \mu) - J^*(\mu)| = 0.$$

Theorem 5.7 implies that to find a near optimal policy for the POMDP, it is sufficient to compute an optimal policy for the finite-state belief-MDP with sufficiently many states, extend this policy to the original state space of the belief-MDP, and then construct the corresponding policy for the POMDP.

In the following, we discuss explicit methods to quantize the set of probability measures on X, that is, the belief space Z.

5.3 Quantization of the Belief Space

An explicit construction for an application requires a properly defined metric on Z. One can metrize the set of probability measures defined on a Borel space under the weak topology using various distance measures. Building on this fact, in the following we present explicit methods for the quantization of the sets of probability measures for the cases where X is finite, a compact subset of a finite dimensional Euclidean space, or the finite-dimensional Euclidean space itself, and $p(\cdot|a)$ is independent of the state variable x.

5.3.1 *Construction with Finite* X

If the state space is finite with $|\mathsf{X}| = m$, then $\mathsf{Z} = \mathcal{P}_v(\mathsf{X}) = \mathcal{P}(\mathsf{X})$, and Z is a simplex in $\mathbb{R}^m$. In this case, Euclidean distance can be used to metrize Z. Indeed, one can make use of the algorithm in [107] (see also [22]) to quantize Z in a nearest neighbor manner. To this end, for each $n \geq 1$, define

$$\mathsf{Z}_n := \left\{ (p_1, \ldots, p_m) \in \mathbb{Q}^m : p_i = \frac{k_i}{n}, \sum_{i=1}^{m} k_i = n \right\}, \tag{5.10}$$

where $\mathbb{Q}$ is the set of rational numbers and $n, k_1, \ldots, k_m \in \mathbb{Z}_+$. The set Z_n is called *type lattice* by analogy with the concept of *types* in information theory [33, Chapter 12]. Then, the algorithm that computes the nearest neighbor levels can be described as follows:

Algorithm Given $z \in \mathsf{Z}$, find nearest $y \in \mathsf{Z}_n$:

(1) Compute values $(i = 1, \ldots, m)$

$$k_i' = \left\lfloor n z_i + \frac{1}{2} \right\rfloor \text{and } n' = \sum_{i=1}^{m} k_i'.$$

(2) If $n' = n$ the nearest y is given by $(\frac{k_1'}{n}, \ldots, \frac{k_m'}{n})$. Otherwise, compute the errors

$$\delta_i = k_i' - n z_i,$$

and sort them

$$\frac{-1}{2} \leq \delta_{i_1} \leq \delta_{i_2} \leq \ldots \leq \delta_{i_m} \leq \frac{1}{2}.$$

(3) Let $\Delta = n' - n$. If $\Delta > 0$, set

$$k_{i_j} = \begin{cases} k'_{i_j} & \text{if } j = 1, \ldots, m - \Delta - 1 \\ k'_{i_j} - 1 & \text{if } j = m - \Delta, \ldots, m. \end{cases}$$

If $\Delta < 0$, set

$$k_{i_j} = \begin{cases} k'_{i_j} + 1 & \text{if } j = 1, \ldots, |\Delta| \\ k'_{i_j} & \text{if } j = |\Delta| + 1, \ldots, m. \end{cases}$$

Then, the nearest y is given by $(\frac{k_1}{n}, \ldots, \frac{k_m}{n})$.

One can also compute the maximum radius of the quantization regions for this algorithm. To this end, let d_∞ and d_p denote respectively the metrics induced by L_∞ and L_p ($p \geq 1$) norms on $\mathbb{R}^m$, which metrizes the weak topology on Z. Then, we have [107, Proposition 2]

$$b_\infty := \max_{z \in \mathsf{Z}} \min_{y \in \mathsf{Z}_n} d_\infty(z, y) = \frac{1}{n}\left(1 - \frac{1}{m}\right),$$

$$b_2 := \max_{z \in \mathsf{Z}} \min_{y \in \mathsf{Z}_n} d_2(z, y) = \frac{1}{n}\sqrt{\frac{a(m-a)}{m}},$$

$$b_1 := \max_{z \in \mathsf{Z}} \min_{y \in \mathsf{Z}_n} d_1(z, y) = \frac{1}{n}\frac{2a(m-a)}{m},$$

where $a = \lfloor m/2 \rfloor$. Hence, for each $n \geq 1$, the set Z_n is an b_j-net in Z with respect to d_j metric, where $j \in \{\infty, 2, 1\}$.

5.3.2 *Construction with Compact* X

The analysis in the previous subsection shows that a finitely supported measure can be approximated through *type lattices*. Thus, if compactly supported probability measures can be approximated with those having finite support, the analysis in Section 5.3.1 yields approximately optimal policies. In the following, we assume that X is a compact subset of $\mathbb{R}^d$ for some $d \geq 1$. Then $\mathsf{Z} := \mathcal{P}_v(\mathsf{X}) = \mathcal{P}(\mathsf{X})$ is also compact (under the weak convergence topology) and can be metrized using the Wasserstein metric W_1. Note that W_1 is originally defined as (see [130, Theorem 6.9]):

$$W_1(\mu, \nu) = \inf_{\eta \in \mathcal{H}(\mu,\nu)} \int_{\mathsf{X}\times\mathsf{X}} \|x - x'\| \eta(dx, dx'),$$

where $\mathcal{H}(\mu, \nu)$ denotes the set of probability measures on $\mathsf{X} \times \mathsf{X}$ with first marginal μ and second marginal ν.

For each $n \geq 1$, let Q_n be some lattice quantizer [58] on X such that $\|x - Q_n(x)\| < 1/n$ for all $x \in \mathsf{X}$. Set $\mathsf{X}_n = Q_n(\mathsf{X})$, i.e., the output levels of Q_n (note that X_n is finite since X is compact). Then, one can approximate any probability measure in Z with probability measures in

$$\mathcal{P}(\mathsf{X}_n) := \Big\{\mu \in \mathcal{P}(\mathsf{X}) : \mu(\mathsf{X}_n) = 1\Big\}.$$

Indeed, for any $\mu \in \mathsf{Z}$, we have [81, Theorem 2.6]

$$\begin{aligned}\inf_{\mu' \in \mathcal{P}(\mathsf{X}_n)} W_1(\mu, \mu') &\leq \inf_{Q:\mathsf{X} \to \mathsf{X}_n} \int_{\mathsf{X}} \|x - Q(x)\| \mu(dx) \\ &\leq \int_{\mathsf{X}} \|x - Q_n(x)\| \mu(dx) \leq \frac{1}{n}.\end{aligned}$$

Once this is obtained, we can further approximate the probability measure induced by Q_n via the algorithm introduced in Section 5.3.1 with asymptotic performance guarantees. Thus, through a sequence of type lattices Z_{m_n} as given in (5.10) with a successively refined support set so that $\mathsf{X}_n \subset \mathsf{X}_{n+1}$ for $n \in \mathbb{N}$ with $m_n = |\mathsf{X}_n|$, one can quantize Z to obtain a sequence of finite-state-action MDPs through (5.9) leading to Theorem 5.7.

For some related properties of approximations of probability measures with those with finite support, and the relation to optimal quantization, we refer the reader to [81].

5.3.3 Construction with Non-compact X

Here we assume that $\mathsf{X} = \mathbb{R}^d$ for some $d \geq 1$ and that Assumption 5.1 holds for $v(x) = \|x\|^2$. In this case, $\mathsf{Z} := \mathcal{P}_v(\mathsf{X})$ becomes the set of probability measures with finite second moment and F_m is the set of probability measures with finite second moments bounded by m. We endow here Z with the bounded-Lipschitz metric ρ_{BL}, which metrizes weak convergence and is defined as [130, p.109]:

$$\rho_{BL}(\mu, \nu) := \sup_{\|f\|_{BL} \leq 1} \left| \int_{\mathsf{X}} f(x)\mu(dx) - \int_{\mathsf{X}} f(x)\nu(dx) \right|, \tag{5.11}$$

where

$$\|f\|_{BL} := \|f\| + \|f\|_{\mathrm{Lip}}.$$

Note that W_1 can be used as an upper bound to ρ_{BL} for non-compact X since W_1 can equivalently be written as (see Section 4.5):

$$W_1(\mu, \nu) := \sup_{\|f\|_{\mathrm{Lip}} \leq 1} \left| \int_{\mathsf{X}} f(x)\mu(dx) - \int_{\mathsf{X}} f(x)\nu(dx) \right|.$$

Comparing this with (5.11), it follows that

$$\rho_{BL} \leq W_1. \tag{5.12}$$

We first describe the discretization procedure for F_m. For each $n \geq 1$, set $K := [-n, n]^d$ and let q_n denote a lattice quantizer on K satisfying

$$\sup_{x \in K} \|x - q_n(x)\| < 1/n.$$

Let K_n denote the set of output levels of q_n; that is, $K_n = q_n(K)$. Define

$$Q_n(x) = \begin{cases} q_n(x) & \text{if } x \in K \\ 0 & \text{if } x \in K^c, \end{cases}$$

Let $\mathsf{X}_n = K_n$. Then, any measure in F_m can be approximated by probability measures in

$$\mathscr{P}(\mathsf{X}_n) := \left\{ \mu \in \mathscr{P}(\mathsf{X}) : \mu(\mathsf{X}_n) = 1 \right\}.$$

Indeed, for any $\mu \in F_m$, we have

$$\begin{aligned} \inf_{\mu' \in \mathscr{P}(\mathsf{X}_n)} \rho_{BL}(\mu, \mu') &\leq \inf_{\mu' \in \mathscr{P}(\mathsf{X}_n)} W_1(\mu, \mu') \qquad (5.13) \\ &\leq \inf_{Q: \mathsf{X} \to \mathsf{X}_n} \int_{\mathsf{X}} \|x - Q(x)\| \mu(dx) \\ &\leq \int_{\mathsf{X}} \|x - Q_n(x)\| \mu(dx) \\ &= \int_{K} \|x - Q_n(x)\| \mu(dx) + \int_{K^c} \|x\| \mu(dx) \\ &\leq \frac{1}{n} + \int_{\{\|x\| > n\}} \|x\|^2 \mu(dx) \frac{1}{n} \\ &\leq \frac{(1 + m)}{n} \qquad (5.14) \end{aligned}$$

In the derivation above, (5.13) follows from (5.12). Thus, μ in F_m can be approximated by the $\mu_n \in \mathscr{P}(\mathsf{X}_n)$, which is induced by the quantizer Q_n, with a bound $\rho_{BL}(\mu, \mu_n) \leq (1+m)/n$. Then, similar to Section 5.3.2, we can further approximate probability measure μ_n via the algorithm introduced in Section 5.3.1 with again asymptotic performance guarantees. Thus, analogous to compact case, using a sequence of type lattices Z_{m_n} as given in (5.10) with a successively refined support set $X_n \subset X_{n+1}$ for $n \in \mathbb{N}$ with $m_n = |\mathsf{X}_n|$, one can quantize Z to obtain a sequence of finite-state-action MDPs through (5.9).

5.3.4 Construction for Special Models Leading to Quantized Beliefs with Continuous Support

So far, we have obtained quantized beliefs where each such quantized belief measure was supported on a finite set. For some applications, this may not be efficient and it may be more desirable to quantize the measurement space appropriately. For some further applications, a parametric representation of the set of allowable beliefs may be available and the construction of bins may be more immediate through quantizing the parameters in a parametric class. What is essential in such models is that the bins designed to construct the finite belief-MDP correspond to balls which are *small* under the metrics that metrize the weak convergence.

Quantized Measures Through Quantized Measurements

For this section, we assume that transition probability $p(\,\cdot\,|a)$ is independent of the state variable x, $\mathsf{Y} \subset \mathbb{R}^p$ for some $p \geq 1$, and Assumption 5.1 holds for some v. In the view of Theorem 5.5, as a preprocessing step, we quantize the action space A. Letting A_q represents the output levels of this quantizer, in the sequel, we assume that the action space is A_q.

Since $\kappa(dx, dy|a) := r(dy|x) \otimes p(dx|a)$, we have

$$\begin{aligned} R(x \in A, y \in B|z, a) &= \int_{\mathsf{X}} \kappa(A, B|a) z(dx') \\ &= \kappa(A, B|a), \end{aligned}$$

and so, the disintegration of R becomes

$$R(dx, dy|a) = H(dy|a) \otimes F(dx|a, y).$$

Then, η is given by

$$\eta(\,\cdot\,|a) = \int_{\mathsf{Y}} \delta_{F(\,\cdot\,|a,y)}(\,\cdot\,)\, H(dy|a).$$

This implies that we can take the following set as the state space Z of the fully observed model instead of $\mathscr{P}_v(\mathsf{X})$:

$$\mathsf{Z} = \biggl\{ F(\,\cdot\,|a, y) : (a, y) \in \mathsf{A}_q \times \mathsf{Y} \biggr\}.$$

We endow Z with the bounded-Lipschitz metric ρ_{BL}. For each $n \geq 1$, set $L := [-n, n]^p$ and let l_n denote a lattice quantizer on L satisfying

$$\sup_{y \in L} \| y - l_n(y) \| < 1/n.$$

Let Y_n denote the set of output levels of l_n; that is, $\mathsf{Y}_n = l_n(L)$. Define

$$q_n(y) = \begin{cases} l_n(y) & \text{if } y \in L \\ 0 & \text{if } y \in L^c. \end{cases}$$

Then, finite set $\mathsf{Z}_n \subset \mathsf{Z}$, which is used to quantize Z, is given by

$$\mathsf{Z}_n = \biggl\{ F(\,\cdot\,|a, y) : (a, y) \in \mathsf{A}_q \times \mathsf{Y}_n \biggr\},$$

and the corresponding quantizer $Q_n : \mathsf{Z} \to \mathsf{Z}_n$ is defined as follows: given $z = F(\,\cdot\,|a, y)$, we define

$$Q_n(z) = F(\,\cdot\,|a, q_n(y)).$$

Note that to use Q_n for constructing finite models, we have to obtain an upper bound on the ρ_{BL}-distance between z and $Q_n(z)$. This can be achieved under various assumptions on the system components. One such assumption is the following: (i) $\mathsf{X} = \mathbb{R}^d$ for some $d \geq 1$, (ii) Y is compact, (iii) $p(dx|a) = g_p(x|a)m(dx)$ and $r(dy|x) = g_r(y|x)m(dy)$, (iv) g_r is Lipschitz continuous with Lipschitz constant K_r, $g_r > \theta$ for some $\theta > 0$, and $\sup_{\{(y,x) \in \mathsf{Y} \times \mathsf{X}\}} |g_r(y, x)| =: \|g_r\| < \infty$. Since Y is compact, there exists $\epsilon(n)$ such that $\epsilon(n) \to 0$ as $n \to \infty$ and $\|y - q_n(y)\| \leq \epsilon(n)$ for all $y \in \mathsf{Y}$. Under the above assumptions, we have

$$F(dx|a, y) = f(x|a, y)m(dx),$$

where

$$f(x|a, y) = \frac{g_r(y|x)g_p(x|a)}{\int_{\mathsf{X}} g_r(y|x)g_p(x|a)m(dx)}.$$

Since the bounded-Lipschitz metric ρ_{BL} is upper bounded by the total variation distance, we obtain

$$\begin{aligned}\rho_{BL}(z, Q_n(z)) &\leq \|z - Q_n(z)\|_{TV} \\ &= \int_{\mathsf{X}} \big|f(x|a, y) - f(x|a, q_n(y))\big| m(dx) \\ &\leq \frac{2\|g_r\| K_r}{\theta^2} \|y - q_n(y)\| \\ &\leq \frac{2\|g_r\| K_r}{\theta^2} \epsilon(n).\end{aligned}$$

Hence, Q_n is a legitimate quantizer for constructing the finite models. Section 5.4.2 exhibits another example where we have such an upper bound.

Construction from a Parametrically Represented Class

For some applications, the set of belief measures can be first approximated by some parametric class of measures, where the parameters belong to some low-dimensional space [25, 26, 155]. For instance, in [155], densities of belief measures are projected onto exponential family of densities using the *Kullback-Leibler (KL) divergence*, where it was assumed that projected beliefs are close enough to true beliefs in terms of cost functions. In [25], densities are parameterized by unimodal Gaussian distributions and the parameterized MDP are solved through a Monte Carlo simulation based method. In [26], densities are represented by sufficient statistics, and in particular represented by Gaussian distributions, and the parameterized MDP is solved through fitted value iteration algorithm. However, among these works, only the [155] develop rigorous error bounds for their algorithms using the KL divergence and others do not specify distance measures to quantify parametric representation approximations.

In these methods, if the parameterized beliefs are sufficient to represent true beliefs as it was shown in [155], then the method presented in the earlier sections (of first quantizing the state space, and then quantizing the beliefs on the state space) may not be necessary and one can, by quantizing the parameters for the class of beliefs considered, directly construct the finite belief-MDP. As noted earlier, what is essential in such methods is that the bins designed to construct the finite belief-MDP correspond to balls which are *small* under the metrics that metrize the weak convergence. This is possible if the projected beliefs are provably close to the true beliefs with respect to some metric that generates the weak topology or with respect to some (pseudo) distance which is stronger than weak topology. For instance, since convergence in KL-divergence is stronger than weak convergence, the projected beliefs constructed in [155] indeed satisfy this requirement. Hence, one can apply our results to conclude the convergence of the reduced model to the original model in [155]. Note that the KL-divergence is a very strong pseudo-distance measure which is even stronger than total variation (by Pinsker's inequality [57]) and for being able to quantize a set of probability measures with finitely many balls as defined by such a distance measure requires very strict assumptions on the allowable beliefs and it in particular requires equicontinuity conditions (see, e.g., [149, Lemma 4.3]). In turn, it is in general necessary to assume that transition probability and observation channel satisfy very strong regularity conditions.

5.4 Numerical Examples

In this section, we consider two examples in order to illustrate our results numerically. Since computing true costs of the policies obtained from the finite models is intractable, we only compute the value functions of the finite models and illustrate their converge as $n \to \infty$. We note that all results in this chapter apply with straightforward modifications for the case of maximizing reward instead of minimizing cost.

5.4.1 Example with Finite X

First we consider a machine repair problem in order to illustrate our results numerically. In this model, we have $\mathsf{X} = \mathsf{A} = \mathsf{Y} = \{0, 1\}$ with the following interpretation:

$$X_t = \begin{cases} 1 & \text{machine is working at time } t \\ 0 & \text{machine is not working at time } t, \end{cases}$$

$$A_t = \begin{cases} 1 & \text{machine is being repaired at time } t \\ 0 & \text{machine is not being repaired at time } t, \end{cases}$$

and

$$Y_t = \begin{cases} 1 & \text{machine is measured to be working at time } t \\ 0 & \text{machine is measured to be not working at time } t. \end{cases}$$

There are two sources of uncertainty in the model. The first one is the measurement uncertainty. The probability that the measured state is not the true state is given by ε; that is,

$$\mathsf{Pr}\{Y_t = 0|X_t = 1\} = \mathsf{Pr}\{Y_t = 1|X_t = 0\} = \varepsilon.$$

In other words, there is a binary symmetric channel with crossover probability ε between the state process and the observation process.

The second uncertainty comes from the repair process. In this case, κ is the probability that the machine repair was successful given an initial "not working" state:

$$\mathsf{Pr}\{X_{t+1} = 1|X_t = 0, A_t = 1\} = \kappa.$$

Finally, the probability that the machine does not break down in one time step is denoted by α:

$$\mathsf{Pr}\{X_{t+1} = 1|X_t = 1, A_t = 0\} = \alpha.$$

The one-stage cost function for this model is given by:

$$c(x,a) = \begin{cases} R + E & x = 0 \text{ and } a = 1 \\ E & x = 0 \text{ and } a = 0 \\ 0 & x = 1 \text{ and } a = 0 \\ R & x = 1 \text{ and } a = 1, \end{cases}$$

where R is defined to be the cost of repair and E is the cost incurred by a broken machine. The cost function to be minimized is the discounted cost function with a discount factor β.

In order to find the approximately optimal policies, we first construct the belief space formulation of the above model. Note that the state space of the belief space model is the interval $[0, 1]$. Hence, we can use uniform quantization on $[0, 1]$ to obtain the finite model.

For the numerical results, we use the following parameters: $\varepsilon = 0.17$, $\kappa = 0.9$, $\alpha = 0.9545$, and $\beta = 0.3$. We selected 20 different values for the number n of grid points to discretize $[0, 1]$: $n = 10, 20, 30, \ldots, 200$. The grid points are chosen uniformly. For each n, the finite-state models are constructed as in Section 5.2.3.

Figure 5.1 shows the graph of the value functions of the finite models corresponding to the different values of n (number of grid points), when the initial state is $x = 1$. It can be seen that the value functions converge (to the value function of the original model by Theorem 4.10).

5.4.2 *Example with Compact* X

In this example we consider the following model:

$$x_{t+1} = \exp\{-\theta_1 a_t + v_t\},\ t = 0, 1, 2, \ldots \tag{5.15}$$

$$y_t = x_t + \xi_t,\ t = 0, 1, 2, \ldots \tag{5.16}$$

where $\theta_1 \in \mathbb{R}_+$, x_t is the state at t, and a_t is the action at t. The one-stage "reward" function is $u(x_t - a_t)$, where u is some utility function. In this model, the goal is to maximize the discounted reward. This model is the modified and partially observed version of the Fisheries Management Model in Section 4.6.2.

The state and action spaces are $\mathsf{X} = \mathsf{A} = [0, L]$, for some $L \in \mathbb{R}_+$, and the observation space is $\mathsf{Y} = [0, K]$ for some $K \in \mathbb{R}_+$. Since θ_1 is a constant, by taking $[0, \frac{L}{\theta_1}]$ as our new action space, instead of the dynamics in equation (5.15) we can write

$$x_{t+1} = \exp\{-a_t + v_t\},\ t = 0, 1, 2 \ldots.$$

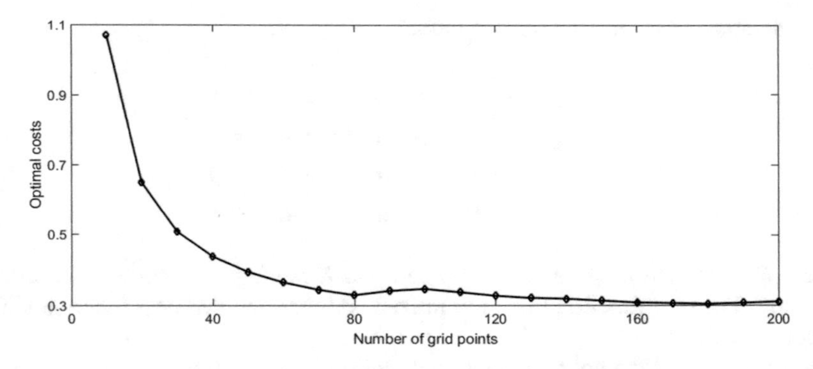

Fig. 5.1 Optimal costs of the finite models when the initial state is $x = 1$

The noise processes $\{v_t\}$ and $\{\xi_t\}$ are sequences of independent and identically distributed (i.i.d.) random variables which have common densities g_v supported on $[0, \lambda]$ and g_ξ supported on $[0, \tau]$, respectively. Therefore, the transition probability $p(\cdot\,|x, a)$ is given by

$$\begin{aligned} p\big(D|x, a\big) &= \Pr\left\{x_{t+1} \in D \,\middle|\, x_t = x, a_t = a\right\} \\ &= \Pr\left\{\exp\{-a + v\} \in D\right\} \\ &= \int_D g_v\big(\log(v) + a\big)\frac{1}{v} m(dv), \end{aligned}$$

for all $D \in \mathscr{B}(\mathbb{R})$ and the observation kernel $r(\,\cdot\,|x)$ is given by

$$r\big(B|x,a\big) = \Pr\left\{y_t \in B \middle| x_t = x\right\}$$
$$= \Pr\left\{x + \xi \in B\right\}$$
$$= \int_B g_\xi(\xi - x)m(d\xi),$$

for all $B \in \mathscr{B}(\mathbb{R})$. To make the model consistent, we must have $\exp\{-a + v\} \in [0, L]$ for all $(a, v) \in [0, L] \times [0, \lambda]$. We assume that g_v and g_ξ are uniform probability density functions; that is, $g_v = \frac{1}{\lambda}$ on $[0, \lambda]$ and $g_\xi = \frac{1}{\tau}$ on $[0, \tau]$. Hence, Assumption 5.1 holds for this model with $v(x) = 1$.

In the view of Theorem 5.5, as a preprocessing setup, we quantize the action space A, where the finite set $\mathsf{A}_q = \{a_1, a_2, \ldots, a_q\}$ represents the output levels of this quantizer with $0 < a_1 < a_2 < \ldots < a_q$. In the remainder of this example we assume that the action space is A_q.

We now obtain the stochastic kernels $H(\,\cdot\,|z, a)$ and $F(\,\cdot\,|z, a, y)$ that describe the transition probability η of the reduced MDP. Indeed, we have

$$H(dy|z, a) = h(y|a)m(dy),$$

where $h(y|a)$ is given by

$$h(y|a) = \int_{\mathsf{X}} g_\xi(y - x)g_v(\log(x) + a)\frac{1}{x}m(dx)$$
$$= \int_{\mathsf{X}} \frac{1}{\tau\lambda} 1_{\left\{\left[y-\tau,y\right]\bigcap\left[\exp\{-a\},\exp\{\lambda-a\}\right]\right\}}(x)\frac{1}{x}m(dx).$$

Similarly, we have

$$F(dx|z, a, y) = f(dx|a, y)m(dx),$$

where $f(x|z, a, y)$ is given by

$$f(x|a, y) = \frac{g_\xi(y - x)g_v(\log(x) + a)\frac{1}{x}}{\int_{\mathsf{X}} g_\xi(y - x)g_v(\log(x) + a)\frac{1}{x}m(dx)}$$
$$= \frac{1_{\left\{\left[y-\tau,y\right]\bigcap\left[\exp\{-a\},\exp\{\lambda-a\}\right]\right\}}(x)\frac{1}{x}}{\int_{\mathsf{X}} 1_{\left\{\left[y-\tau,y\right]\bigcap\left[\exp\{-a\},\exp\{\lambda-a\}\right]\right\}}(x)\frac{1}{x}m(dx)}. \tag{5.17}$$

Hence, for any (z, a), the transition probability $\eta(\,\cdot\,|z, a)$ has a support on the set of probability measures on X having densities given by (5.17). This implies that we can take the following set as the state space Z of the fully observed model instead of $\mathscr{P}(\mathsf{X})$:

$$\mathsf{Z} = \biggl\{ f(x|a, y)m(dx) : (a, y) \in \mathsf{A} \times \mathsf{Y} \text{ and } f \text{ as in (5.17)} \biggr\}.$$

Note that for some $(a, y) \in \mathsf{A} \times \mathsf{Y}$, probability density function $f(x|a, y)$ is not well defined as $[y - \tau, y] \cap [\exp\{-a\}, \exp\{\lambda - a\}] = \emptyset$, and so, we disregard these points. For the rest of the points in $\mathsf{A} \times \mathsf{Y}$, a typical $f(x|a, y)$ can be in the following forms:

$$f(x|a, y) = \frac{1_{[\exp\{-a\}, y]}(x)\frac{1}{x}}{\log(y) + \log(a)} \tag{5.18}$$

$$f(x|a, y) = \frac{1_{[y-\tau, y]}(x)\frac{1}{x}}{\log(y) - \log(y - \tau)} \tag{5.19}$$

$$f(x|a, y) = \frac{1_{[y-\tau, \exp\{\lambda - a\}]}(x)\frac{1}{x}}{\log(\lambda - a) - \log(y - \tau)}. \tag{5.20}$$

For each n, let q_n denote the uniform quantizer on Y having n output levels; that is,

$$q_n : \mathsf{Y} \to \{y_1, \ldots, y_n\} =: \mathsf{Y}_n \subset \mathsf{Y},$$

where $y_j = (j - \frac{1}{2})\Delta_n$, $j = 1, \ldots, n$, and

$$q_n^{-1}(y_j) = \biggl[y_j - \frac{\Delta_n}{2}, y_j + \frac{\Delta_n}{2} \biggr),$$

where $\Delta_n = \frac{K}{n}$. We define

$$\mathsf{Z}_n := \biggl\{ f(x|a, y)m(dx) \in \mathsf{Z} : (a, y) \in \mathsf{A}_q \times \mathsf{Y}_n \biggr\}.$$

Then, the quantizer $Q_n : \mathsf{Z} \to \mathsf{Z}_n$, which is used to construct the finite model, is defined as follows: given $z = f(x|a, y)m(dx)$, we define

$$Q_n(z) = f(x|a, q_n(y))m(dx).$$

To be able to use Q_n for constructing finite models, we need to obtain an upper bound on the ρ_{BL}-distance between z and $Q_n(z)$. To this end, let $\theta > 0$ be a small constant such that $\exp\{-a_q\} - \theta > 0$.

Suppose that the density of z is in the form of (5.19); that is, $y - \tau > \exp\{-a\}$ and $y < \exp\{\lambda - a\}$ for some $a \in \mathsf{A}_q$. Let $y_n := q_n(y)$, $\gamma := \frac{1}{\log(y)-\log(y-\tau)}$, and $\gamma_n := \frac{1}{\log(y_n)-\log(y_n-\tau)}$. Since $\frac{1}{\log(y)-\log(y-\tau)}$ is a continuous function of y and $[\exp\{-a_q\}+\tau, \exp\{\lambda - a_1\}]$ is compact, there exist $\epsilon(\Delta_n)$ which are independent of z such that $|\gamma - \gamma_n| < \epsilon(\Delta_n)$ and $\epsilon(\Delta_n) \to 0$ as $n \to \infty$. We also suppose that $y_n < y$ without loss of generality. Then, for sufficiently large n, by using the inequality $\log(x) \leq x - 1$, we obtain

$$
\begin{aligned}
\rho_{BL}(z, Q_n(z)) &\leq \|z - Q_n(z)\|_{TV} \\
&= \int_{\mathsf{X}} \big| f(x|a, y) - f(x|a, q_n(y)) \big| m(dx) \\
&= \int_{\mathsf{X}} \left| 1_{[y-\tau, y]}(x) \frac{1}{x}\gamma - 1_{[y_n-\tau, y_n]}(x) \frac{1}{x}\gamma_n \right| m(dx) \\
&= \int_{y_n-\tau}^{y-\tau} \frac{1}{x}\gamma_n m(dx) + \int_{y-\tau}^{y_n} \left| \frac{1}{x}\gamma - \frac{1}{x}\gamma_n \right| m(dx) + \int_{y_n}^{y} \frac{1}{x}\gamma m(dx) \\
&= \gamma_n \log(\frac{y-\tau}{y_n-\tau}) + |\gamma - \gamma_n| \log(\frac{y_n}{y-\tau}) + \gamma \log(\frac{y}{y_n}) \\
&\leq 2K_1 \frac{\Delta_n}{K_2} + L_1 \epsilon(\Delta_n),
\end{aligned}
$$

where

$$
\begin{aligned}
K_1 &:= \frac{1}{\log(\frac{\exp\{\lambda - a_1\}+\tau}{\exp\{\lambda - a_1\}})} \\
K_2 &:= \exp\{-a_q\} - \theta \\
L_1 &:= \log(\frac{\exp\{-a_q\}+\tau}{\exp\{-a_q\}}).
\end{aligned}
$$

Hence, $\rho_{BL}(z, Q_n(z)) \to 0$ as $n \to \infty$. Similar computations can be done for $z \in \mathsf{Z}$ of the form (5.18) and (5.20). This implies that Q_n is a legitimate quantizer to construct finite-state models.

For the numerical results, we use the following values of the parameters:

$$
\lambda = 1, \ \tau = 0.5, \ \beta = 0.2.
$$

The utility function u is taken to be quadratic function; i.e., $u(t) = t^2$. As a preprocessing setup, we first uniformly discretize the action space A by using the 20 grid points. Then, we selected 99 different values for the number n of grid points to discretize the state space Z using the quantizer Q_n, where n varies from 29 to 1436.

We use the value iteration algorithm to compute the value functions of the finite models. The simulation was implemented by using MATLAB. Figure 5.2 displays the graph of these value functions corresponding to the different values for the number of grid points, when the initial state is $x = 2$. The figure illustrates that the value functions of the finite models converge (to the value function of the original model by Theorem 4.10).

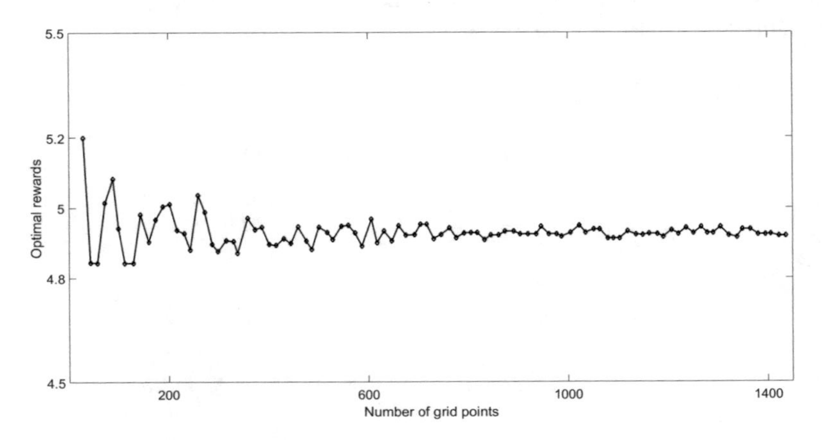

Fig. 5.2 Optimal rewards of the finite models when the initial state is $x = 2$

5.5 Concluding Remarks

In this chapter, the approximation of discrete-time partially observed Markov decision processes was studied. An essential observation was that establishing strong continuity properties for the reduced (belief) model is quite difficult for general state and action models, whereas weak continuity can be established under fairly mild conditions on the transition kernel for the original model and the measurement equations. Thus, the results obtained in *Chapters 3* and *4* under weak continuity conditions allowed for the applicability to partially observed models.

There exist various prior results for obtaining approximately optimal policies for POMDPs; however, the prior literature has been mostly numerical and computational, and in general has studied finite state and action spaces as it is fairly difficult to establish regularity conditions that should be satisfied by the transition probability of the belief-MDP so that the convergence is guaranteed for general state space models. Reference [102] develops a computational algorithm, utilizing structural convexity properties of the value function of belief-MDPs, for the solutions of POMDPs when the state space is continuous and action and measurements are discrete, and with further extensions to continuous action and measurements. Reference [131] provides an algorithm which may be regarded as a quantization of the belief space. However, no rigorous convergence results are given regarding this algorithm. References [124] and [101] present quantization based algorithms for the belief state, where the state, measurement, and the action sets are finite. References [154] and [155] provide an explicit quantization method for the set of probability measures containing the belief states, where the state space, unlike in many other contributions in the literature, is continuous. The quantization is done through the approximations as measured by Kullback-Leibler divergence. In [147] the authors consider the near optimality of finite-state controllers that are finite-state probabilistic automatons taking observations as inputs and producing controls as the outputs. A special case for these type of controllers are the ones that only use finite observation history. A similar finite memory approximation is developed in [133]. In [146] the authors establish finite-state approximation schemes for the belief-MDPs under both discounted cost and average cost criteria using concavity properties of the corresponding value function and show that approximate costs can be used as lower bounds for the optimal cost function. A similar finite-state approximation is considered in [89, 156] using concavity and convexity properties of the value function for the discounted cost criterion. In [36] the approximation problem for partially observed optimal stopping problems is approached by applying a two-step discretization procedure based on optimal quantization. First, the state space is discretized by quantizing an underlying reference measure and then the resulting approximate filter and the observation variable are jointly discretized. This procedure yields an approximation of the value function with explicit error bounds. We refer the reader to the survey papers [90, 132] and the books [82, 111] for further algorithmic and computational procedures for approximating POMDPs. Some further related results were discussed in Section 5.3.4.

As we noted earlier, an essential component for the analysis in this chapter is that the bins used to construct the finite belief-MDP correspond to balls which are *small* under the metric that metrizes the topology of weak convergence on the set of probability measures. Many contributions in the literature provide constructions which indeed conform with this principle and the analysis here can be viewed as a theoretical justification for such procedures.

Chapter 6
Approximations for Constrained Markov Decision Problems

This chapter studies the finite-state approximation of a discrete-time constrained Markov decision process with compact state space, under the discounted and average cost criteria. Using the linear programming formulation of the constrained discounted problem, we prove the convergence of the optimal value function of the finite-state model to the optimal value function of the original model. Under further continuity conditions on the transition probability of the original discounted model, we also establish a method to compute approximately optimal policies. For the average cost criterion, instead of using the finite-state linear programming approximation method, we use a direct method to establish analogous results under drift and minorization conditions which guarantee the geometric ergodicity of Markov chains induced by stationary policies.

6.1 Introduction

Although approximate dynamic programming is an efficient method to compute nearly optimal policies, it cannot be applied to constrained stochastic control problems as the dynamic programming principle does not in general hold in the presence of constraints. Therefore, many approximation methods for unconstrained MDPs cannot be applied directly to constrained MDPs.

In this chapter, we develop a finite-state linear programming approximation approach for constrained discounted-cost Markov decision processes. We first prove that the value function of the finite-state model converges to the value function of the original model. Then, we present a method to compute approximately optimal policies for the original problem under additional continuity assumptions on the transition probability. For the average cost, instead of using finite-state linear programming approximation, we use a direct method. To this end, we impose drift and minorization conditions, introduced in Section 4.2.2, on the components of

N. Saldi et al., *Finite Approximations in Discrete-Time Stochastic Control*, Systems & Control: Foundations & Applications,
https://doi.org/10.1007/978-3-319-79033-6_6

the control model. Recall that these conditions guarantee that the Markov chains induced by stationary policies satisfy a geometric ergodicity condition; that is, the t-step transition probabilities of the Markov chain converges geometrically to the invariant distribution. Similar to the discounted case, we first show the convergence of the value function of the finite-state model to the value function of the original model and then introduce a method for computing approximately optimal policies.

We do not consider the quantitative behavior of the approximation; that is, we do not establish explicit error bounds for the approximation error. Because of this, the conditions on the transition probability and the one-stage cost function are quite mild. In addition, we also present a method to compute approximately optimal policies for the original problem via finite-state models. The significance of this result lies in the fact that using well-known algorithms for finite-state constrained MDPs, one can compute approximately optimal policies for the original problem. Therefore, the approach is constructive in that it yields approximately optimal policies.

6.2 Constrained Markov Decision Processes

We consider a discrete-time constrained Markov decision process (MDP) with components

$$\big(\mathsf{X}, \mathsf{A}, p, c, \gamma, \boldsymbol{d}, \boldsymbol{k}\big),$$

where $\gamma \in \mathscr{P}(\mathsf{X})$ denotes the initial distribution. Here, the components $(\mathsf{X}, \mathsf{A}, p, c)$ define a usual unconstrained Markov decision process. The last two components are the vectors of costs $\boldsymbol{d} = (d_1, \ldots, d_q) : \mathsf{X} \times \mathsf{A} \to \mathbb{R}^q_+$ and constraints $\boldsymbol{k} = (k_1, \ldots, k_q) \in \mathbb{R}^q_+$, that will be used to define the constraints of the problem. Note that the history spaces and the set of policies are defined as in an unconstrained MDP.

For each policy $\pi \in \Pi$ and $\beta \in (0, 1)$ consider the β-discounted cost functions

$$J(\pi, \gamma) = (1-\beta)\mathbb{E}^{\pi}_{\gamma}\bigg[\sum_{t=0}^{\infty} \beta^t c(X_t, A_t)\bigg],$$

$$J_l(\pi, \gamma) = (1-\beta)\mathbb{E}^{\pi}_{\gamma}\bigg[\sum_{t=0}^{\infty} \beta^t d_l(X_t, A_t)\bigg] \text{ for } l = 1, \ldots, q.$$

We normalize the usual discounted cost by the coefficient $(1 - \beta)$ to simplify some technical details. Similarly, for each policy $\pi \in \Pi$, consider the average cost functions

$$V(\pi,\gamma) = \limsup_{T\to\infty} \frac{1}{T} \mathbb{E}_{\gamma}^{\pi}\biggl[\sum_{t=0}^{T-1} c(X_t,A_t)\biggr],$$

$$V_l(\pi,\gamma) = \limsup_{T\to\infty} \frac{1}{T} \mathbb{E}_{\gamma}^{\pi}\biggl[\sum_{t=0}^{T-1} d_l(X_t,A_t)\biggr] \text{ for } l=1,\ldots,q.$$

Using the above notation, the constrained decision problems for discounted and average cost criteria can be defined as follows:

$$\textbf{(CP)} \text{ minimize } W(\pi,\gamma)$$
$$\text{subject to } W_l(\pi,\gamma) \leq k_l \text{ for } l=1,\ldots,q, \; \pi \in \Pi,$$

where $W \in \{J, V\}$. In this chapter, we assume that the following conditions hold for both discounted cost and average cost criteria.

Assumption 6.1

(a) X *and* A *are compact.*
(b) The one-stage cost function c *and the constraint functions* d_l $(l = 1,\ldots,q)$ *are continuous.*

6.2.1 Finite-State Model

Recall the finite-state model MDP_n constructed in Section 4.2 with the following components: X_n is the state space, A is the action space, p_n is the transition probability, c_n is the one-stage cost function,

$$\gamma_n := Q_n * \gamma$$

is the initial distribution. Furthermore, for each n, define the functions $\boldsymbol{d}_n = (d_{1,n},\ldots,d_{q,n}) : \mathsf{X}_n \times \mathsf{A} \to \mathbb{R}_+^q$ by

$$d_{l,n}(x_{n,i},a) := \int_{\mathscr{S}_{n,i}} d_l(x,a)\nu_{n,i}(dx) \text{ for } l=1,\ldots,q.$$

Then, with a slight abuse of notation, we define MDP_n as the constrained Markov decision process with components

$$\bigl(\mathsf{X}_n, \mathsf{A}, p_n, c_n, \gamma_n, \boldsymbol{d}_n, \boldsymbol{k}\bigr).$$

History spaces, policies, cost, and constraint functions are defined analogously to the original model. Let Π_n, Φ_n, and $\mathbb{F}_n$ denote the set of all, randomized stationary and deterministic stationary policies for MDP_n, respectively. For each policy $\pi \in \Pi_n$, the β-discounted and average costs for the functions c_n and $d_{l,n}$ ($l = 1, \ldots, q$) are defined analogously. Then, as in the definition of **(CP)**, we define the constrained problem for MDP_n by

$$(\mathbf{CP}_n) \text{ minimize } W_n(\pi, \gamma_n)$$
$$\text{subject to } W_{l,n}(\pi, \gamma_n) \leq k_l \; (l = 1, \ldots, q), \; \pi \in \Pi_n,$$

where $W \in \{J, V\}$.

6.3 Approximation of Discounted Cost Problems

We first consider the approximation of the value function; that is, we show that the value function of the finite-state model converges to the value function of the original model. Then, we establish a method for computing near optimal policies for the original constrained Markov decision process. It is important to note that in the results that we are aware of (e.g., [41, 43]), dealing with the approximation of constrained MDPs, only the convergence of the value functions is established. In this section, we also establish the convergence (in terms of costs) of the optimal policies of the finite-state models to the optimal policy of the original one.

6.3.1 Approximation of Value Function

The analysis in this section is nonconstructive, the goal being establishing the convergence of the value function of the approximate model. The conditions here are weaker than those in the next section, where a constructive approach is developed for computing near optimal policies. Specifically, in this section we prove that the value function of $(\mathbf{CP}_n)$ converges to the value function of **(CP)**, i.e.,

$$\inf(\mathbf{CP}_n) \to \inf(\mathbf{CP}) \tag{6.1}$$

as $n \to \infty$. To prove (6.1), the following assumptions will be imposed; additional assumptions will be made for the problem of computing near optimal policies in Section 6.3.2.

Assumption 6.2 *We assume that Assumption 6.1 holds. In addition, we assume*

(a) The stochastic kernel $p(\cdot|x, a)$ is weakly continuous in (x, a) and setwise continuous in a.
(b) There exists a policy $\pi \in \Pi$ such that $J_l(\pi, \gamma) < k_l$ for $l = 1, \ldots, q$. In other words, $(J_1(\pi, \gamma), \ldots, J_q(\pi, \gamma)) + \boldsymbol{\alpha}' = \mathbf{k}$ for some $\boldsymbol{\alpha}' > \mathbf{0}$.

Before proving (6.1), we formulate both (**CP**) and ($\mathbf{CP}_n$) as linear programs on appropriate linear spaces. The duals of these linear programs will play a key role in proving (6.1). We refer the reader to [67] and [68, Chapter 6] for a linear programming formulation of constrained MDPs with discounted cost function.

Consider the vector spaces $\big(\mathcal{M}(\mathsf{X} \times \mathsf{A}), B(\mathsf{X} \times \mathsf{A})\big)$ and $\big(\mathcal{M}(\mathsf{X}), B(\mathsf{X})\big)$. Let us define bilinear forms on $\big(\mathcal{M}(\mathsf{X} \times \mathsf{A}), B(\mathsf{X} \times \mathsf{A})\big)$ and on $\big(\mathcal{M}(\mathsf{X}), B(\mathsf{X})\big)$ by letting

$$\langle \zeta, v \rangle := \int_{\mathsf{X}\times\mathsf{A}} v(x,a)\zeta(dx,da),$$

$$\langle \mu, u \rangle := \int_{\mathsf{X}} u(x)\mu(dx).$$

These bilinear forms constitute dualities between $\mathcal{M}(\mathsf{X}\times\mathsf{A})$, $\mathcal{M}(\mathsf{X})$, and $B(\mathsf{X}\times\mathsf{A})$, $B(\mathsf{X})$, respectively [11, Chapter IV.3]. Hence, the topologies on these spaces should be understood as the weak topology of the duality induced by these bilinear forms. For any $\zeta \in \mathcal{M}(\mathsf{X} \times \mathsf{A})$, let $\hat{\zeta} \in \mathcal{M}(\mathsf{X})$ denote the marginal of ζ on X, i.e.,

$$\hat{\zeta}(\,\cdot\,) = \zeta(\,\cdot\, \times \mathsf{A}).$$

We define the linear maps $T : \mathcal{M}(\mathsf{X} \times \mathsf{A}) \rightarrow \mathcal{M}(\mathsf{X})$ and $L : \mathcal{M}(\mathsf{X} \times \mathsf{A}) \times \mathbb{R}^q \rightarrow \mathcal{M}(\mathsf{X}) \times \mathbb{R}^q$ by

$$T\zeta(\,\cdot\,) = \hat{\zeta}(\,\cdot\,) - \beta \int_{\mathsf{X}\times\mathsf{A}} p(\,\cdot\,|x,a)\zeta(dx,da)$$

$$L(\zeta, \boldsymbol{\alpha}) = \Big(T\zeta, \langle \zeta, \boldsymbol{d} \rangle + \boldsymbol{\alpha}\Big),$$

where $\langle \zeta, \boldsymbol{d} \rangle := \big(\langle \zeta, d_1 \rangle, \ldots, \langle \zeta, d_q \rangle\big)$. Then, (**CP**) is equivalent to the following equality constrained linear program [67, Lemma 3.3 and Section 4], which is also denoted by (**CP**):

$$\begin{aligned} &\textbf{(CP)} \text{ minimize } \langle (\zeta, \boldsymbol{\alpha}), (c, \mathbf{0}) \rangle = \langle \zeta, c \rangle \\ &\qquad \text{subject to } L(\zeta, \boldsymbol{\alpha}) = \big((1-\beta)\gamma, \boldsymbol{k}\big), \\ &\qquad (\zeta, \boldsymbol{\alpha}) \in \mathcal{M}_+(\mathsf{X} \times \mathsf{A}) \times \mathbb{R}^q_+. \end{aligned} \tag{6.2}$$

Indeed, for any policy $\pi \in \Pi$, define the (expected) occupation measure as

$$\zeta^{\pi}(C) := (1-\beta) \sum_{t=0}^{\infty} \beta^t P^{\pi}_{\gamma}\Big[(X_t, A_t) \in C\Big], \; C \in \mathcal{B}(\mathsf{X} \times \mathsf{A}).$$

Note that ζ^{π} is a probability measure on $\mathsf{X}\times\mathsf{A}$ as a result of the normalizing constant $(1-\beta)$. In the absence of this normalization, we would have to deal with nonprobability measures which complicates the analysis. One can prove that ζ^{π} satisfies

$$\hat{\zeta}^{\pi}(\cdot) = (1-\beta)\gamma(\cdot) + \beta \int_{\mathsf{X}\times\mathsf{A}} p(\cdot\,|x,a)\zeta^{\pi}(dx,da). \tag{6.3}$$

Conversely, if any finite measure ζ satisfies (6.3), then it is an occupation measure of some policy $\pi \in \Phi$. Using the occupation measure, we can write

$$J(\pi,\gamma) = \langle \zeta^{\pi}, c\rangle \text{ and } J_l(\pi,\gamma) = \langle \zeta^{\pi}, d_l\rangle.$$

One can prove that the set of randomized stationary policies Φ is sufficient for optimality in (**CP**). Therefore, (**CP**) can be written in the following alternative form:

$$\begin{aligned}
\textbf{(CP)}\ &\text{minimize } \langle \zeta, c\rangle \\
&\text{subject to } \langle \zeta, d_l\rangle \leq k_l \text{ for } l = 1,\ldots,q \\
&\hat{\zeta}(\cdot) = (1-\beta)\gamma(\cdot) + \beta \int_{\mathsf{X}\times\mathsf{A}} p(\cdot\,|x,a)\zeta(dx,da), \\
&\zeta \in \mathcal{M}_+(\mathsf{X}\times\mathsf{A}).
\end{aligned}$$

From this alternative formulation, it is straightforward to obtain (6.2).

The adjoint $T^* : B(\mathsf{X}) \to B(\mathsf{X}\times\mathsf{A})$ of T and the adjoint $L^* : B(\mathsf{X}) \times \mathbb{R}^q \to B(\mathsf{X}\times\mathsf{A}) \times \mathbb{R}^q$ of L are given by (see [67, Section 4])

$$T^*u(x,a) = u(x) - \beta \int_{\mathsf{X}} u(y)p(dy|x,a),$$

$$L^*(u,\boldsymbol{\delta}) = \biggl(T^*u + \sum_{l=1}^{q} \delta_l d_l, \boldsymbol{\delta}\biggr).$$

Hence, the dual (**CP***) of (**CP**) is

$$\begin{aligned}
\textbf{(CP*)}\ &\text{maximize } \bigl\langle \bigl((1-\beta)\gamma, \boldsymbol{k}\bigr), (u,\boldsymbol{\delta})\bigr\rangle = (1-\beta)\langle \gamma, u\rangle + \langle \boldsymbol{k}, \boldsymbol{\delta}\rangle \\
&\text{subject to } L^*(u,\boldsymbol{\delta}) \leq (c,\boldsymbol{0}),\ (u,\boldsymbol{\delta}) \in B(\mathsf{X}) \times \mathbb{R}^q.
\end{aligned}$$

Here, $L^*(u,\boldsymbol{\delta}) \leq (c,\boldsymbol{0})$ can be written more explicitly as

$$u(x) \leq c(x,a) - \sum_{l=1}^{q} \delta_l d_l(x,a) + \beta \int_{\mathsf{X}} u(y)p(dy|x,a)$$

for all $(x,a) \in \mathsf{X}\times\mathsf{A}$, where $\delta_l \leq 0$ for $l = 1,\ldots,q$.

By replacing $(\mathsf{X}, p, c, \boldsymbol{d}, \gamma)$ with $(\mathsf{X}_n, p_n, c_n, \boldsymbol{d}_n, \gamma_n)$ above, we can write the equivalent equality constraint linear program for (**CP**$_n$) as follows:

$$
\begin{aligned}
(\mathbf{CP}_n)\ & \text{minimize } \langle \zeta, c_n \rangle \\
& \text{subject to } L_n(\zeta, \boldsymbol{\alpha}) = \big((1-\beta)\gamma_n, \boldsymbol{k}\big), \qquad (6.4)\\
& (\zeta, \boldsymbol{\alpha}) \in \mathcal{M}_+(\mathsf{X}_n \times \mathsf{A}) \times \mathbb{R}^q_+,
\end{aligned}
$$

where $T_n : \mathcal{M}(\mathsf{X}_n \times \mathsf{A}) \to \mathcal{M}(\mathsf{X}_n)$ and $L_n : \mathcal{M}(\mathsf{X}_n \times \mathsf{A}) \times \mathbb{R}^q \to \mathcal{M}(\mathsf{X}_n) \times \mathbb{R}^q$ are given by

$$
T_n\zeta(\,\cdot\,) = \hat{\zeta}(\,\cdot\,) - \beta \sum_{i=1}^{k_n} \int_{\mathsf{A}} p_n(\,\cdot\,|x_{n,i}, a)\zeta(x_{n,i}, da)
$$

$$
L_n(\zeta, \boldsymbol{\alpha}) = \bigg(T_n\zeta, \langle \zeta, \boldsymbol{d}_n \rangle + \boldsymbol{\alpha}\bigg).
$$

Similarly, the dual $(\mathbf{CP}_n^*)$ of $(\mathbf{CP}_n)$ is given by

$$
\begin{aligned}
(\mathbf{CP}_n^*)\ & \text{maximize } (1-\beta)\langle \gamma_n, u \rangle + \langle \boldsymbol{k}, \boldsymbol{\delta} \rangle \\
& \text{subject to } L_n^*(u, \boldsymbol{\delta}) \le (c_n, \mathbf{0}),\ \ (u, \boldsymbol{\delta}) \in B(\mathsf{X}_n) \times \mathbb{R}^q,
\end{aligned}
$$

where the adjoint $T_n^* : B(\mathsf{X}_n) \to B(\mathsf{X}_n \times \mathsf{A})$ of T_n and the adjoint $L_n^* : B(\mathsf{X}_n) \times \mathbb{R}^q \to B(\mathsf{X}_n \times \mathsf{A}) \times \mathbb{R}^q$ of L_n are given by

$$
T_n^* u(x_{n,i}, a) = u(x_{n,i}) - \beta \sum_{j=1}^{k_n} u(x_{n,j}) p_n(x_{n,j}|x_{n,i}, a),
$$

$$
L_n^*(u, \boldsymbol{\delta}) = \bigg(T_n^* u + \sum_{l=1}^{q} \delta_l d_{l,n}, \boldsymbol{\delta}\bigg).
$$

Here, $L_n^*(u, \boldsymbol{\delta}) \le (c_n, \mathbf{0})$ can be written more explicitly as

$$
u(x_{n,i}) \le c_n(x_{n,i}, a) - \sum_{l=1}^{q} \delta_l d_{l,n}(x_{n,i}, a) + \beta \sum_{j=1}^{k_n} u(x_{n,j}) p_n(x_{n,j}|x_{n,i}, a)
$$

for all $(x_{n,i}, a) \in \mathsf{X}_n \times \mathsf{A}$, where $\delta_l \le 0$ for $l = 1, \ldots, q$.

If Assumption 6.2 holds, then by [67, Theorems 3.2 and 4.3] we have

$$
\begin{aligned}
\sup(\mathbf{CP}^*) &= \min(\mathbf{CP}) \\
\sup(\mathbf{CP}_n^*) &= \min(\mathbf{CP}_n),
\end{aligned}
$$

where the "min" notation signifies that there exist optimal policies for $(\mathbf{CP})$ and $(\mathbf{CP}_n)$. Furthermore, if $(\zeta^*, \boldsymbol{\alpha})$ and $(\zeta_n^*, \boldsymbol{\alpha}_n)$ are minimizers for $(\mathbf{CP})$ and $(\mathbf{CP}_n)$, respectively, then the optimal (randomized stationary) policies $\varphi^* \in \Phi$ and $\varphi_n^* \in \Phi_n$ for MDP and MDP$_n$ are given by disintegrating ζ^* and ζ_n^* as

$$\zeta^*(dx, da) = \varphi^*(da|x)\hat{\zeta}^*(dx),$$

$$\zeta_n^*(dx, da) = \varphi_n^*(da|x)\hat{\zeta}_n^*(dx).$$

The following theorem is the main result of this section.

Theorem 6.1 *We have*

$$\lim_{n\to\infty} \left| \min(\mathbf{CP}_n) - \min(\boldsymbol{CP}) \right| = 0; \tag{6.5}$$

that is, the value function of $(\mathbf{CP}_n)$ *converges to the value function of* $(\mathbf{CP})$ *as* $n \to \infty$.

To prove Theorem 6.1, for each $n \geq 1$ we introduce, another constrained MDP, denoted by $\overline{\text{MDP}}_n$, with the components

$$\big(\mathsf{X}, \mathsf{A}, q_n, b_n, \boldsymbol{r}_n, \gamma\big),$$

where the definitions of the stochastic kernel q_n and the one-stage cost function b_n are given in Section 4.2.2, and $\boldsymbol{r}_n = (r_{1,n}, \ldots, r_{q,n}) : \mathsf{X} \times \mathsf{A} \to \mathbb{R}_+^q$ is defined as

$$r_{l,n}(x, a) = \int d_l(z, a)\nu_{n,i_n(x)}(dz).$$

Recall that $i_n : \mathsf{X} \to \{1, \ldots, k_n\}$ maps x to the index of the quantization region it belongs to. As before, the constrained decision problem that corresponds to $\overline{\text{MDP}}_n$ can be formulated as an equality constrained linear program given by

$$\begin{aligned} (\overline{\mathbf{CP}}_n) \text{ minimize } & \langle \zeta, b_n \rangle \\ \text{subject to } & \overline{L}_n(\zeta, \boldsymbol{\alpha}) = \big((1-\beta)\gamma, \boldsymbol{k}\big), \\ & (\zeta, \boldsymbol{\alpha}) \in \mathcal{M}_+(\mathsf{X} \times \mathsf{A}) \times \mathbb{R}_+^q, \end{aligned}$$

where $\overline{T}_n : \mathcal{M}(\mathsf{X} \times \mathsf{A}) \to \mathcal{M}(\mathsf{X})$ and $\overline{L}_n : \mathcal{M}(\mathsf{X} \times \mathsf{A}) \times \mathbb{R}^q \to \mathcal{M}(\mathsf{X}) \times \mathbb{R}^q$ are given by

$$\overline{T}_n\zeta(\,\cdot\,) = \hat{\zeta}(\,\cdot\,) - \beta \int_{\mathsf{X}\times\mathsf{A}} q_n(\,\cdot\,|x, a)\zeta(dx, da)$$

$$\overline{L}_n(\zeta, \boldsymbol{\alpha}) = \left(\overline{T}_n\zeta, \langle \zeta, \boldsymbol{r}_n \rangle + \boldsymbol{\alpha}\right).$$

Furthermore, the dual $(\overline{\mathbf{CP}}_n^*)$ of $(\overline{\mathbf{CP}}_n)$ is given by

$$(\overline{\mathbf{CP}}_n^*) \text{ maximize } (1-\beta)\langle \gamma, u\rangle + \langle \boldsymbol{k}, \boldsymbol{\delta}\rangle$$
$$\text{subject to } \overline{L}_n^*(u,\boldsymbol{\delta}) \leq (b_n, \mathbf{0}),\ (u,\boldsymbol{\delta}) \in B(\mathsf{X}_n) \times \mathbb{R}^q,$$

where the adjoint $\overline{T}_n^* : B(\mathsf{X}) \to B(\mathsf{X} \times \mathsf{A})$ of $\overline{T}_n$ and the adjoint $\overline{L}_n^* : B(\mathsf{X}) \times \mathbb{R}^q \to B(\mathsf{X} \times \mathsf{A}) \times \mathbb{R}^q$ of $\overline{L}_n$ are as follows:

$$\overline{T}_n^* u(x,a) = u(x) - \beta \int_{\mathsf{X}} u(y) q_n(dy|x,a),$$

$$\overline{L}_n^*(u,\boldsymbol{\delta}) = \biggl(\overline{T}_n^* u + \sum_{l=1}^{q} \delta_l r_{l,n}, \boldsymbol{\delta}\biggr).$$

Here, $\overline{L}_n^*(u,\boldsymbol{\delta}) \leq (b_n, \mathbf{0})$ can be written more explicitly as

$$u(x) \leq b_n(x,a) - \sum_{l=1}^{q} \delta_l r_{l,n}(x,a) + \beta \int_{\mathsf{X}} u(y) q_n(dy|x,a)$$

for all $(x,a) \in \mathsf{X} \times \mathsf{A}$, where $\delta_l \leq 0$ for $l = 1, \ldots, q$. The following result states that MDP_n and $\overline{\mathrm{MDP}}_n$ are essentially identical.

Lemma 6.2 *We have*

$$\min(\boldsymbol{CP}_n) = \min(\overline{\boldsymbol{CP}}_n), \tag{6.6}$$

and if the randomized stationary policy $\varphi^* \in \Phi_n$ is optimal for $(\boldsymbol{CP}_n)$, then its extension $\overline{\varphi}^*$ to X is also optimal for $(\overline{\boldsymbol{CP}}_n)$ with the same cost function, where $\overline{\varphi}^*$ is given by $\overline{\varphi}^*(\,\cdot\,|x) = \varphi^*(\,\cdot\,|Q_n(x))$.

Proof We first prove that any policy $\overline{\varphi} \in \Phi$, which is an extension (to X) of a feasible policy $\varphi \in \Phi_n$ for $(\mathbf{CP}_n)$, is also feasible for $(\overline{\mathbf{CP}}_n)$; that is, it satisfies the constraints in $(\overline{\mathbf{CP}}_n)$.

Fix any $\varphi \in \Phi_n$ feasible for $(\mathbf{CP}_n)$ and extend φ to X by letting $\overline{\varphi}(\,\cdot\,|x) = \varphi(\,\cdot\,|Q_n(x))$. Let $\zeta \in \mathcal{M}_+(\mathsf{X}_n \times \mathsf{A})$ denote the expected *occupation measure* of φ [67, p. 274], which can be disintegrated as $\zeta(dx,da) = \varphi(da|x)\hat{\zeta}(dx) =: \varphi \otimes \hat{\zeta}$. Hence, ζ satisfies

$$T_n \zeta = T_n\bigl(\varphi \otimes \hat{\zeta}\bigr) = (1-\beta)\gamma_n \text{ and } \langle \zeta, \boldsymbol{d}_n \rangle \leq \boldsymbol{k}. \tag{6.7}$$

Let $\zeta_e = \overline{\varphi} \otimes \hat{\zeta}_e \in \mathcal{M}_+(\mathsf{X} \times \mathsf{A})$ denote the expected occupation measure corresponding to $\overline{\varphi}$. Hence, ζ_e satisfies $\overline{T}_n \zeta_e = (1-\beta)\gamma$, or more explicitly

$$(1-\beta)\gamma(\cdot)=\hat{\zeta}_e(\cdot)-\beta\int_{\mathsf{X}\times\mathsf{A}} q_n(\cdot|x,a)\overline{\varphi}(da|x)\hat{\zeta}_e(dx). \tag{6.8}$$

Note that $q_n(\cdot|x,a)=q_n(\cdot|y,a)$ and $\overline{\varphi}(\cdot|x)=\overline{\varphi}(\cdot|y)$ if x,y are in the same partition. Hence, if we take the pushforward of (6.8) with respect to Q_n, we obtain

$$\begin{aligned}(1-\beta)\gamma_n &= Q_n*\hat{\zeta}_e-\beta\int_{\mathsf{X}\times\mathsf{A}} Q_n*q_n(\cdot|x,a)\overline{\varphi}(da|x)\hat{\zeta}_e(dx)\\ &= Q_n*\hat{\zeta}_e-\beta\sum_{i=1}^{k_n}\int_{\mathsf{A}} p_n(\cdot|x_{n,i},a)\overline{\varphi}(da|x_{n,i})Q_n*\hat{\zeta}_e(x_{n,i})\\ &= T_n\big(\varphi\otimes Q_n*\hat{\zeta}_e\big).\end{aligned}$$

This and (6.7) imply that $Q_n*\hat{\zeta}_e=\hat{\zeta}$. Thus, we have

$$\langle\zeta_e,\boldsymbol{r}_n\rangle=\langle\zeta,\boldsymbol{d}_n\rangle\leq\boldsymbol{k}, \tag{6.9}$$

and

$$\langle\zeta_e,b_n\rangle=\langle\zeta,c_n\rangle, \tag{6.10}$$

where (6.9) states that $\overline{\varphi}$ is feasible for ($\overline{\mathbf{CP}}_n$), and (6.10) states that cost functions of φ and $\overline{\varphi}$ are the same.

Therefore, to prove the lemma, it is enough to prove

$$\sup(\mathbf{CP}_n^*)=\sup(\overline{\mathbf{CP}}_n^*),$$

since $\sup(\mathbf{CP}_n^*)=\min(\mathbf{CP}_n)\geq\inf(\overline{\mathbf{CP}}_n)\geq\sup(\overline{\mathbf{CP}}_n^*)$, where the first inequality follows from what we have just proved above.

For each $\boldsymbol{\delta}\in\mathbb{R}^q$, define $\Gamma_n^{\boldsymbol{\delta}}: B(\mathsf{X}_n)\to B(\mathsf{X}_n)$ and $\overline{\Gamma}_n^{\boldsymbol{\delta}}: B(\mathsf{X})\to B(\mathsf{X})$ by

$$\begin{aligned}\Gamma_n^{\boldsymbol{\delta}}u(x_{n,i}) &= \min_{a\in\mathsf{A}}\int\left[c_{\boldsymbol{\delta}}(z,a)+\beta\int_{\mathsf{X}}\hat{u}(y)p(dy|x,a)\right]\nu_{n,i}(dz),\\ \overline{\Gamma}_n^{\boldsymbol{\delta}}u(x) &= \min_{a\in\mathsf{A}}\int\left[c_{\boldsymbol{\delta}}(z,a)+\beta\int_{\mathsf{X}}u(y)p(dy|x,a)\right]\nu_{n,i_n(x)}(dz),\end{aligned}$$

where $c_{\boldsymbol{\delta}}(z,a)=c(z,a)-\sum_{l=1}^q\delta_l d_l(z,a)$ and $\hat{u}=u\circ Q_n$ for $u\in B(\mathsf{X}_n)$. Here, $\Gamma_n^{\boldsymbol{\delta}}$ is the *Bellman optimality operator* of the unconstrained Markov decision process with components $\big(\mathsf{X}_n,\mathsf{A},p_n,c_n-\sum_{l=1}^q\delta_l d_{l,n},\gamma_n\big)$, and $\overline{\Gamma}_n^{\boldsymbol{\delta}}$ is the Bellman optimality operator of the unconstrained Markov decision process with components

$\big(\mathsf{X}, \mathsf{A}, q_n, b_n - \sum_{l=1}^{q} \delta_l r_{l,n}, \gamma\big)$. They are both contractions with modulus β. Furthermore, the fixed point $\overline{u}_{n,\delta}^*$ of $\overline{\Gamma}_n^{\delta}$ is the piecewise constant extension of the fixed point $u_{n,\delta}^*$ of Γ_n^{δ}, i.e.,

$$\overline{u}_{n,\delta}^* = u_{n,\delta}^* \circ Q_n.$$

Using these operators one can rewrite $(\mathbf{CP}_n^*)$ and $(\overline{\mathbf{CP}}_n^*)$ as

$$(\mathbf{CP}_n^*) \text{ maximize } (1-\beta)\langle \gamma_n, u\rangle + \langle \boldsymbol{k}, \boldsymbol{\delta}\rangle$$
$$\text{subject to } u \leq \Gamma_n^{\delta} u,\ \ \boldsymbol{\delta} \in \mathbb{R}_-^q,\ \ u \in B(\mathsf{X}_n),$$

$$(\overline{\mathbf{CP}}_n^*) \text{ maximize } (1-\beta)\langle \gamma, u\rangle + \langle \boldsymbol{k}, \boldsymbol{\delta}\rangle$$
$$\text{subject to } u \leq \overline{\Gamma}_n^{\delta} u,\ \ \boldsymbol{\delta} \in \mathbb{R}_-^q,\ \ u \in B(\mathsf{X}_n).$$

Observe that if $u \leq \Gamma_n^{\delta} u$, then $u \leq u_{n,\delta}^*$. Indeed, for any $m \geq 1$, we have $u \leq \big(\Gamma_n^{\delta}\big)^m u$ since Γ_n^{δ} is monotone $\big(\text{i.e., } u \leq v \text{ implies } \Gamma_n^{\delta} u \leq \Gamma_n^{\delta} v\big)$. Furthermore, $\big(\Gamma_n^{\delta}\big)^m u$ converges to $u_{n,\delta}^*$ by the Banach fixed point theorem. Hence, $u \leq u_{n,\delta}^*$. The same conclusion can be made for $\overline{u}_{n,\delta}^*$. Thus, we can write

$$(\mathbf{CP}_n^*) \text{ maximize } (1-\beta)\langle \gamma_n, u_{n,\delta}^*\rangle + \langle \boldsymbol{k}, \boldsymbol{\delta}\rangle$$
$$\text{subject to } \boldsymbol{\delta} \in \mathbb{R}_-^q,$$

$$(\overline{\mathbf{CP}}_n^*) \text{ maximize } (1-\beta)\langle \gamma, \overline{u}_{n,\delta}^*\rangle + \langle \boldsymbol{k}, \boldsymbol{\delta}\rangle$$
$$\text{subject to } \boldsymbol{\delta} \in \mathbb{R}_-^q.$$

Since $\overline{u}_{n,\delta}^* = u_{n,\delta}^* \circ Q_n$, we have

$$\langle \gamma, \overline{u}_{n,\delta}^*\rangle = \langle \gamma_n, u_{n,\delta}^*\rangle,$$

for all $\boldsymbol{\delta} \in \mathbb{R}_-^q$. Thus,

$$\sup(\mathbf{CP}_n^*) = \sup(\overline{\mathbf{CP}}_n^*),$$

which completes the proof of the lemma. □

Lemma 6.2 implies that to prove Theorem 6.1, it is sufficient to show that

$$\lim_{n\to\infty} \big| \min(\overline{\mathbf{CP}}_n) - \min(\mathbf{CP}) \big| = 0.$$

We use this fact in the proof of Theorem 6.1.

For each $\boldsymbol{\delta} \in \mathbb{R}^q$, we define, analogous to Γ_n^{δ} and $\overline{\Gamma}_n^{\delta}$, the operator $\Gamma^{\delta} : B(\mathsf{X}) \to B(\mathsf{X})$ by

$$\Gamma^{\delta} u(x) = \min_{a\in\mathsf{A}} \left[c(x,a) - \sum_{l=1}^{q} \delta_l d_l(x,a) + \beta \int_{\mathsf{X}} u(y) p(dy|x,a) \right].$$

Here Γ^{δ} is the Bellman optimality operator for the unconstrained Markov decision process with the components $\big(\mathsf{X}, \mathsf{A}, p, c - \sum_{l=1}^{q} \delta_l d_l, \gamma\big)$, and moreover, it is a contraction operator with modulus β. Let u_{δ}^* denote the fixed point of Γ^{δ}. Similar to the proof of Lemma 6.2, one can rewrite $(\mathbf{CP}^*)$ as

$$(\mathbf{CP}^*) \text{ maximize } (1-\beta)\langle \gamma, u_{\delta}^* \rangle + \langle \boldsymbol{k}, \boldsymbol{\delta} \rangle$$
$$\text{subject to } \boldsymbol{\delta} \in \mathbb{R}_-^q.$$

Let us define functions $G_n : \mathbb{R}_-^q \to \mathbb{R}$ and $G : \mathbb{R}_-^q \to \mathbb{R}$ by

$$G_n(\boldsymbol{\delta}) = (1-\beta)\langle \gamma, \overline{u}_{n,\delta}^* \rangle + \langle \boldsymbol{k}, \boldsymbol{\delta} \rangle$$
$$G(\boldsymbol{\delta}) = (1-\beta)\langle \gamma, u_{\delta}^* \rangle + \langle \boldsymbol{k}, \boldsymbol{\delta} \rangle.$$

Hence,

$$\sup(\overline{\mathbf{CP}}_n^*) = \sup_{\boldsymbol{\delta}\in\mathbb{R}_-^q} G_n(\boldsymbol{\delta}) \text{ and } \sup(\mathbf{CP}^*) = \sup_{\boldsymbol{\delta}\in\mathbb{R}_-^q} G(\boldsymbol{\delta}).$$

By Theorem 4.10, we have $\lim_{n\to\infty} \|\overline{u}_{n,\delta}^* - u_{\delta}^*\| = 0$ for all $\boldsymbol{\delta} \in \mathbb{R}_-^q$. This implies that

$$G_n(\boldsymbol{\delta}) \to G(\boldsymbol{\delta}) \tag{6.11}$$

as $n \to \infty$ for all $\boldsymbol{\delta} \in \mathbb{R}_-^q$.

Lemma 6.3 *There exists $n(\boldsymbol{k}) \geq 1$ such that for each $n \geq n(\boldsymbol{k})$, one can find $(\zeta_n, \boldsymbol{\alpha}_n) \in \mathcal{M}_+(\mathsf{X} \times \mathsf{A}) \times \mathbb{R}_+^q$ feasible for $(\overline{\mathbf{CP}}_n)$ and $\boldsymbol{\alpha}_n \geq \boldsymbol{\alpha}'/2$, where $\boldsymbol{\alpha}'$ is the vector in Assumption 6.2-(d).*

Proof Let $\mathscr{P}$ and $\mathscr{P}_n$ denote the set of expected occupation measures for MDP and $\overline{\text{MDP}}_n$, respectively, i.e.,

$$\mathscr{P} = \big\{\zeta : T\zeta = (1-\beta)\gamma\big\},$$
$$\mathscr{P}_n = \big\{\zeta : T_n\zeta = (1-\beta)\gamma\big\}.$$

It can be proved that for each $l = 1, \ldots, q$, we have

$$\inf_{\zeta \in \mathscr{P}} \langle \zeta, d_l \rangle = (1-\beta)\langle J_l^*, \gamma \rangle,$$
$$\inf_{\zeta \in \mathscr{P}_n} \langle \zeta, r_{l,n} \rangle = (1-\beta)\langle J_{l,n}^*, \gamma \rangle,$$

where

$$J_l^*(x) = \inf_{\pi \in \Pi} (1-\beta)\mathbb{E}_x^{\pi}\bigg[\sum_{t=0}^{\infty} \beta^t d_l(X_t, A_t)\bigg],$$

$$J_{l,n}^*(x) = \inf_{\pi \in \Pi} (1-\beta)\mathbb{E}_x^{\pi}\bigg[\sum_{t=0}^{\infty} \beta^t r_{l,n}(X_t, A_t)\bigg].$$

By Theorem 4.10, we have $\|J_{l,n}^* - J_l^*\| \to 0$ as $n \to \infty$, and therefore, $|\inf_{\zeta\in\mathscr{P}_n}\langle \zeta, r_{l,n}\rangle - \inf_{\zeta\in\mathscr{P}}\langle \zeta, d_l\rangle| \to 0$ as $n \to \infty$. By Assumption 6.2-(d), we have

$$\inf_{\zeta \in \mathscr{P}} \langle \zeta, d_l \rangle + \alpha_l' \leq k_l \text{ for } l = 1, \ldots, q.$$

Thus, one can find $n(\boldsymbol{k}) \geq 1$ large enough such that

$$\inf_{\zeta \in \mathscr{P}_n} \langle \zeta, r_{l,n} \rangle + \frac{\alpha_l'}{2} \leq k_l \text{ for } l = 1, \ldots, q.$$

This completes the proof. □

Proposition 6.4 *There exist* $\boldsymbol{\delta}_n^* \in \mathbb{R}_-^q$ $\big(n \geq n(\boldsymbol{k})\big)$ *and* $\boldsymbol{\delta}^* \in \mathbb{R}_-^q$ *such that*

$$G_n(\boldsymbol{\delta}_n^*) = \sup(\overline{\boldsymbol{CP}}_n^*),$$
$$G(\boldsymbol{\delta}^*) = \sup(\boldsymbol{CP}^*),$$

and $\|\boldsymbol{\delta}_n^*\|_1, \|\boldsymbol{\delta}^*\|_1 \leq K < \infty$, *where* $\|\boldsymbol{\delta}\|_1 := \sum_{l=1}^q |\delta_l|$.

Proof Although the proposition can be deduced from [41, Theorems 3.6 and 4.10], for the sake of completeness we give the proof here (following the arguments in [41]). The first part will be proved only for $(\overline{\mathbf{CP}}_n^*)$ since the proof for $(\mathbf{CP}^*)$ be the same.

Fix any $n \geq n(\boldsymbol{k})$. By Lemma 6.3, there exists $(\zeta_n, \boldsymbol{\alpha}_n)$ feasible for $(\overline{\mathbf{CP}}_n^*)$ such that $\boldsymbol{\alpha}_n \geq \boldsymbol{\alpha}'/2$. Let

$$\mathscr{C}_n = \bigcup_{\zeta \in \mathscr{P}_n} \left\{ \boldsymbol{m} \in \mathbb{R}^q : \langle \zeta, \boldsymbol{r}_n \rangle + \boldsymbol{\alpha} = \boldsymbol{m} \text{ for some } \boldsymbol{\alpha} \in \mathbb{R}_+^q \right\}$$

and define the function $\mathscr{V}_n : \mathscr{C}_n \to \mathbb{R}$ by

$$\mathscr{V}_n(\boldsymbol{m}) = \min \left\{ \langle \zeta, b_n \rangle : \zeta \in \mathscr{P}_n \text{ and } \langle \zeta, \boldsymbol{r}_n \rangle + \boldsymbol{\alpha} = \boldsymbol{m}, \ \boldsymbol{\alpha} \in \mathbb{R}_+^q \right\}.$$

Note that $\mathscr{V}_n(\boldsymbol{k}) = \min(\overline{\mathbf{CP}}_n)$. It can be proved that $\mathscr{C}_n$ is a convex subset of $\mathbb{R}^q$ and $\mathscr{V}_n$ is a convex function. Since $\boldsymbol{k} \in \operatorname{int} \mathscr{C}_n$, by [2, Theorem 7.12] there exists $\boldsymbol{\delta}_n^* \in \mathbb{R}^q$ such that for all $\boldsymbol{m} \in \mathscr{C}_n$

$$\mathscr{V}_n(\boldsymbol{m}) - \mathscr{V}_n(\boldsymbol{k}) \geq -\langle \boldsymbol{\delta}_n^*, \boldsymbol{m} - \boldsymbol{k} \rangle,$$

i.e., $\boldsymbol{\delta}_n^*$ is a subgradient of $\mathscr{V}_n$ at $\boldsymbol{k}$. Since $\mathscr{V}_n(\boldsymbol{m}) \leq \mathscr{V}_n(\boldsymbol{k})$ when $\boldsymbol{m} \geq \boldsymbol{k}$, we have $\boldsymbol{\delta}_n^* \geq \mathbf{0}$.

For any $\zeta \in \mathscr{P}_n$, we have $\langle \zeta, \boldsymbol{r}_n \rangle \in \mathscr{C}_n$ with $\boldsymbol{\alpha} = \mathbf{0}$ and $\mathscr{V}_n(\langle \zeta, \boldsymbol{r}_n \rangle) \leq \langle \zeta, b_n \rangle$. Hence,

$$\langle \zeta, b_n \rangle - \mathscr{V}_n(\boldsymbol{k}) \geq -\big\langle \boldsymbol{\delta}_n^*, \langle \zeta, \boldsymbol{r}_n \rangle - \boldsymbol{k} \big\rangle$$

for all $\zeta \in \mathscr{P}_n$ and therefore,

$$\begin{aligned} \min(\overline{\mathbf{CP}}_n) = \mathscr{V}_n(\boldsymbol{k}) &\leq \inf_{\zeta \in \mathscr{P}_n} \big\langle \zeta, b_n + \langle \boldsymbol{\delta}_n^*, \boldsymbol{r}_n \rangle \big\rangle - \langle \boldsymbol{\delta}_n^*, \boldsymbol{k} \rangle \\ &= (1-\beta)\langle \gamma, \overline{u}_{n,-\boldsymbol{\delta}_n^*}^* \rangle + \langle -\boldsymbol{\delta}_n^*, \boldsymbol{k} \rangle \\ &\leq \sup(\overline{\mathbf{CP}}_n^*) = \min(\overline{\mathbf{CP}}_n). \end{aligned}$$

Thus, $(\overline{u}_{n,-\boldsymbol{\delta}_n^*}^*, -\boldsymbol{\delta}_n^*)$ is the maximizer for $(\overline{\mathbf{CP}}_n^*)$. The same argument for $(\mathbf{CP}_n^*)$ proves the existence of a maximizer $(u_{-\boldsymbol{\delta}^*}^*, -\boldsymbol{\delta}^*)$. This completes the proof of the first part.

To prove the second part, let $\alpha_n := \min_{l=1,\ldots,q} \alpha_{n,l} \geq \frac{1}{2} \min_{l=1,\ldots,q} \alpha_l' =: \alpha/2 > 0$. Hence, $\boldsymbol{k} - \alpha_n \mathbf{1} \in \mathscr{C}_n$ and therefore,

$$\begin{aligned} \mathscr{V}_n(\boldsymbol{k} - \alpha_n \mathbf{1}) - \mathscr{V}_n(\boldsymbol{k}) \geq -\langle \boldsymbol{\delta}_n^*, -\alpha_n \mathbf{1} \rangle &= \alpha_n \|\boldsymbol{\delta}_n^*\|_1 \\ &\geq \frac{\alpha}{2} \|\boldsymbol{\delta}_n^*\|_1. \end{aligned}$$

Since $\mathscr{V}_n(\boldsymbol{k} - \alpha_n \mathbf{1}) \leq \sup_{\zeta \in \mathscr{P}_n} \langle \zeta, b_n \rangle \leq \|b_n\| = \|c\|$ and $\mathscr{V}_n(\boldsymbol{k}) \geq 0$, we have

$$\|\boldsymbol{\delta}_n^*\|_1 \leq \frac{2\|c\|}{\alpha}.$$

Hence, $\|\boldsymbol{\delta}_n^*\|_1, \|\boldsymbol{\delta}^*\|_1 \leq \max\left\{\|\boldsymbol{\delta}^*\|_1, \frac{2\|c\|}{\alpha}\right\} =: K$. This completes the proof of the second part. □

Proof of Theorem 6.1 Recall that to prove Theorem 6.1, it is sufficient to prove

$$\lim_{n\to\infty} \left| \min(\overline{\mathbf{CP}}_n) - \min(\overline{\mathbf{CP}}) \right| = 0,$$

or equivalently

$$\lim_{n\to\infty} \left| \max(\overline{\mathbf{CP}}_n^*) - \max(\overline{\mathbf{CP}}^*) \right| = 0$$

since there is no duality gap. By Proposition 6.4, the latter equation can be written as

$$\lim_{n\to\infty} \left| \sup_{\boldsymbol{\delta}\in\mathscr{K}} G_n(\boldsymbol{\delta}) - \sup_{\boldsymbol{\delta}\in\mathscr{K}} G(\boldsymbol{\delta}) \right| = 0, \tag{6.12}$$

where $\mathscr{K} = \{\boldsymbol{\delta} \in \mathbb{R}_-^q : \|\boldsymbol{\delta}\|_1 \leq K\}$ is a compact subset of $\mathbb{R}_-^q$. Hence, if we can show that G_n converges to G uniformly on $\mathscr{K}$, then the proof is complete. We prove this by showing the relative compactness of $\{G_n\}$ with respect to the topology of uniform convergence.

First, we note that $\{G_n\}$ is equicontinuous with respect to the metric induced by the norm $\|\cdot\|_1$. Indeed, for any $\boldsymbol{\delta}, \boldsymbol{\delta}'$, we have

$$\begin{aligned}
|G_n(\boldsymbol{\delta}) - G_n(\boldsymbol{\delta}')| &= \left|(1-\beta)\langle \gamma, \overline{u}_{n,\boldsymbol{\delta}}^* \rangle + \langle \boldsymbol{k}, \boldsymbol{\delta} \rangle - (1-\beta)\langle \gamma, \overline{u}_{n,\boldsymbol{\delta}'}^* \rangle - \langle \boldsymbol{k}, \boldsymbol{\delta}' \rangle\right| \\
&\leq (1-\beta)\left|\langle \gamma, \overline{u}_{n,\boldsymbol{\delta}}^* \rangle - \langle \gamma, \overline{u}_{n,\boldsymbol{\delta}'}^* \rangle\right| + \left|\langle \boldsymbol{k}, \boldsymbol{\delta} \rangle - \langle \boldsymbol{k}, \boldsymbol{\delta}' \rangle\right| \\
&= (1-\beta)\Bigg| \inf_{\zeta\in\mathscr{P}_n} \langle \zeta, b_n - \sum_{l=1}^q \delta_l r_{l,n} \rangle \\
&\qquad - \inf_{\zeta\in\mathscr{P}_n} \langle \zeta, b_n - \sum_{l=1}^q \delta_l' r_{l,n} \rangle \Bigg| + \sum_{l=1}^q k_l |\delta_l - \delta_l'| \\
&\leq (1-\beta) \sup_{\zeta\in\mathscr{P}_n} \left| \langle \zeta, \sum_{l=1}^q (\delta_l - \delta_l') r_{l,n} \rangle \right| + \max_{l=1,\ldots,q} k_l \, \|\boldsymbol{\delta} - \boldsymbol{\delta}'\|_1 \\
&\leq \left((1-\beta) \max_{l=1,\ldots,q} \|r_{l,n}\| + \max_{l=1,\ldots,q} k_l \right) \|\boldsymbol{\delta} - \boldsymbol{\delta}'\|_1.
\end{aligned}$$

Since $\|r_{l,n}\| \leq \|d_l\|$, we have $G_n \in \mathrm{Lip}(\mathscr{K}, M)$ for all n, where

$$M = \big((1-\beta) \max_{l=1,\ldots,q} \|d_l\| + \max_{l=1,\ldots,q} k_l\big)$$

and $\mathrm{Lip}(\mathscr{K}, M)$ denotes the set of all Lipschitz continuous functions on $\mathscr{K}$ with Lipschitz constant M. Hence, $\{G_n\}$ is equicontinuous. Furthermore, it is also straightforward to prove that for any $\boldsymbol{\delta} \in \mathscr{K}$, $\{G_n(\boldsymbol{\delta})\}$ is bounded. Thus, by the Arzela-Ascoli theorem, $\{G_n\}$ is relatively compact with respect to the topology of uniform convergence. Recall that $G_n \to G$ pointwise $\big($see (6.11)$\big)$, and therefore, every uniformly convergent subsequence of $\{G_n\}$ must converge to G. Together with the relative compactness of $\{G_n\}$, this implies that G_n converges to G uniformly. This completes the proof. □

6.3.2 Approximation of Optimal Policy

In this section we establish a method for computing near optimal policies for the constrained Markov decision problem (**CP**). To this end, we need to slightly strengthen Assumption 6.2 by replacing Assumption 6.2-(a) with the continuity of $p(\,\cdot\,|x,a)$ in (x,a) with respect to the total variation distance. In this section, we assume that the following assumptions hold:

Assumption 6.3
We suppose that Assumption 6.1 and Assumption 6.2-(b) hold. Furthermore, we assume

(b') The stochastic kernel $p(\,\cdot\,|x,a)$ is continuous in (x,a) with respect to total variation distance.

Remark 6.5 If we view the stochastic kernel $p(\,\cdot\,|x,a)$ as a mapping from $\mathsf{X} \times \mathsf{A}$ to $\mathscr{P}(\mathsf{X})$, then Assumption 6.3 means that this mapping is continuous (here $\mathscr{P}(\mathsf{X})$ is equipped with the total variation distance metric), and therefore uniformly continuous since $\mathsf{X} \times \mathsf{A}$ is compact.

For any $g \in B(\mathsf{X} \times \mathsf{A})$ and any $\pi \in \Pi$, define $J^g(\pi,\gamma)$ and $J_n^g(\pi,\gamma)$ as the β-discounted costs of MDP and $\overline{\mathrm{MDP}}_n$, respectively, when the one-stage cost function is g. For each n, let the randomized stationary policy $\overline{\varphi}_n \in \Phi$ denote the extension of a policy $\varphi_n \in \Phi_n$ to X, i.e.,

$$\overline{\varphi}_n(\,\cdot\,|x) = \varphi_n(\,\cdot\,|Q_n(x)),$$

and therefore,

$$\overline{\varphi}_n(\,\cdot\,|x) = \overline{\varphi}_n(\,\cdot\,|y) \;\text{ if }\; Q_n(x) = Q_n(y). \tag{6.13}$$

If we apply $\overline{\varphi}_n$ both to MDP and $\overline{\text{MDP}}_n$, we obtain two Markov chains, describing the state processes, with the following transition probabilities:

$$P_n(\cdot\,|x) := p(\cdot\,|x, \overline{\varphi}_n) = \int_{\mathsf{A}} p(\cdot\,|x, a)\overline{\varphi}_n(da|x),$$

$$R_n(\cdot\,|x) := q_n(\cdot\,|x, \overline{\varphi}_n) = \int_{\mathsf{A}} q_n(\cdot\,|x, a)\overline{\varphi}_n(da|x).$$

Furthermore, by (6.13) we can write $R_n(\cdot\,|x)$ as

$$R_n(\cdot\,|x) = \int P_n(\cdot\,|z)\nu_{n,i_n(x)}(dz).$$

For any $t \geq 1$, we write $P_n^t(\cdot\,|\gamma)$ and $R_n^t(\cdot\,|\gamma)$ to denote the t-step transition probability of the Markov chains given the initial distribution γ; that is, $P_n^0(\cdot\,|\gamma) = R_n^0(\cdot\,|\gamma) = \gamma$ and for $t \geq 1$

$$P_n^{t+1}(\cdot\,|\gamma) = \int_{\mathsf{X}} P_n(\cdot\,|x)P_n^t(dx|\gamma),$$

$$R_n^{t+1}(\cdot\,|\gamma) = \int_{\mathsf{X}} R_n(\cdot\,|x)R_n^t(dx|\gamma).$$

Before stating the next lemma, we need some new notation. For any $g : \mathsf{X} \times \mathsf{A} \to \mathbb{R}$ and $n \geq 1$, let $g_{\overline{\varphi}_n}(x)$ denote $\int_{\mathsf{A}} g(x, a)\overline{\varphi}_n(da|x)$ and let $g_n(x, a)$ denote $\int_{\mathcal{S}_{n,i_n(x)}} g(z, a)\nu_{n,i_n(x)}(dz)$.

Lemma 6.6 *Let $\{\varphi_n\}$ be a sequence such that $\varphi_n \in \Phi_n$ for all n. Then, for any $g \in \mathscr{C}_b(\mathsf{X} \times \mathsf{A})$ and for any $t \geq 1$, we have*

$$\left| \int_{\mathsf{X}} g_{n,\overline{\varphi}_n}(x)R_n^t(dx|\gamma) - \int_{\mathsf{X}} g_{\overline{\varphi}_n}(x)P_n^t(dx|\gamma) \right| \to 0$$

as $n \to \infty$.

Proof By induction we first prove that for any $t \geq 1$,

$$\lim_{n\to\infty} \left\| R_n^t(\cdot\,|\gamma) - P_n^t(\cdot\,|\gamma) \right\|_{TV} = 0. \tag{6.14}$$

For $t = 1$ the claim holds by the following argument:

$$\begin{aligned}\left\| R_n(\cdot\,|\gamma) - P_n(\cdot\,|\gamma) \right\|_{TV} &\leq \int_{\mathsf{X}} \left\| R_n(\cdot\,|x) - P_n(\cdot\,|x) \right\|_{TV} \gamma(dx) \\ &\leq \int_{\mathsf{X}} \int \left\| P_n(\cdot\,|z) - P_n(\cdot\,|x) \right\|_{TV} \nu_{n,i_n(x)}(dz)\gamma(dx)\end{aligned}$$

$$\leq \int_{\mathsf{X}\times\mathsf{A}} \int \left\| p(\cdot\,|z,a) - p(\cdot\,|x,a) \right\|_{TV} \nu_{n,i_n(x)}(dz) \overline{\varphi}_n(da|x)\gamma(dx)$$

$$\left(\text{since } \overline{\varphi}_n(da|z) = \overline{\varphi}_n(da|x),\ z \in \mathscr{S}_{n,i_n(x)}\right)$$

$$\leq \sup_{(x,a)\in\mathsf{X}\times\mathsf{A}} \sup_{z\in\mathscr{S}_{n,i_n(x)}} \left\| p(\cdot\,|z,a) - p(\cdot\,|x,a) \right\|_{TV}.$$

As the mapping $p(\cdot\,|x,a) : \mathsf{X} \times \mathsf{A} \to \mathscr{P}(\mathsf{X})$ is uniformly continuous (see Remark 6.5), the result follows. Assume the claim is true for $t \geq 1$. Then we have

$$\begin{aligned}
\left\| R_n^{t+1}(\cdot\,|\gamma) - P_n^{t+1}(\cdot\,|\gamma) \right\|_{TV} &= \left\| \int_{\mathsf{X}} R_n(\cdot\,|x) R_n^t(dx|\gamma) - \int_{\mathsf{X}} P_n(\cdot\,|x) P_n^t(dx|\gamma) \right\|_{TV} \\
&\leq \left\| \int_{\mathsf{X}} R_n(\cdot\,|x) R_n^t(dx|\gamma) - \int_{\mathsf{X}} P_n(\cdot\,|x) R_n^t(dx|\gamma) \right\|_{TV} \\
&\quad + \left\| \int_{\mathsf{X}} P_n(\cdot\,|x) R_n^t(dx|\gamma) - \int_{\mathsf{X}} P_n(\cdot\,|x) P_n^t(dx|\gamma) \right\|_{TV} \\
&\leq \int_{\mathsf{X}} \left\| R_n(\cdot\,|x) - P_n(\cdot\,|x) \right\|_{TV} R_n^t(dx|\gamma) \\
&\quad + 2\left\| R_n^t(\cdot\,|\gamma) - P_n^t(\cdot\,|\gamma) \right\|_{TV} \\
&\leq \sup_{(x,a)\in\mathsf{X}\times\mathsf{A}} \sup_{z\in\mathscr{S}_{n,i_n(x)}} \left\| p(\cdot\,|z,a) - p(\cdot\,|x,a) \right\|_{TV} \\
&\quad + 2\left\| R_n^t(\cdot\,|\gamma) - P_n^t(\cdot\,|\gamma) \right\|_{TV}.
\end{aligned}$$

Since the mapping $p(\cdot\,|x,a) : \mathsf{X} \times \mathsf{A} \to \mathscr{P}(\mathsf{X})$ is uniformly continuous, the first term converges to zero. The second term also converges to zero since the claim holds for t. This completes the proof of (6.14).

Using (6.14), we obtain

$$\begin{aligned}
&\left| \int_{\mathsf{X}} g_{n,\overline{\varphi}_n}(x) R_n^t(dx|\gamma) - \int_{\mathsf{X}} g_{\overline{\varphi}_n}(x) P_n^t(dx|\gamma) \right| \\
&\leq \left| \int_{\mathsf{X}} g_{n,\overline{\varphi}_n}(x) R_n^t(dx|\gamma) - \int_{\mathsf{X}} g_{\overline{\varphi}_n}(x) R_n^t(dx|\gamma) \right| \\
&\quad + \left| \int_{\mathsf{X}} g_{\overline{\varphi}_n}(x) R_n^t(dx|\gamma) - \int_{\mathsf{X}} g_{\overline{\varphi}_n}(x) P_n^t(dx|\gamma) \right| \\
&\leq \|g_n - g\| + \|g\| \left\| R_n^t(\cdot\|\gamma) - P_n^t(\cdot\|\gamma) \right\|_{TV}.
\end{aligned}$$

As $g_n \to g$ uniformly and $\left\| R_n^t(\cdot\,|\gamma) - P_n^t(\cdot\,|\gamma) \right\|_{TV} \to 0$, the result follows. □

Proposition 6.7 *For any* $g \in C_b(\mathsf{X} \times \mathsf{A})$, *we have*

$$\lim_{n\to\infty} \left| J_n^{g_n}(\overline{\varphi}_n, \gamma) - J^g(\overline{\varphi}_n, \gamma) \right| = 0$$

as $n \to \infty$.

Proof We have

$$\begin{aligned}
&\limsup_{n\to\infty} \left| J_n^{g_n}(\overline{\varphi}_n, \gamma) - J^g(\overline{\varphi}_n, \gamma) \right| \\
&= (1-\beta) \limsup_{n\to\infty} \left| \sum_{t=0}^{\infty} \beta^t \int_{\mathsf{X}} g_{n,\overline{\varphi}_n}(x) R_n^t(dx|\gamma) \right. \\
&\qquad \left. - \sum_{t=0}^{\infty} \beta^t \int_{\mathsf{X}} g_{\overline{\varphi}_n}(x) P_n^t(dx|\gamma) \right| \\
&\leq (1-\beta) \Bigg(\limsup_{n\to\infty} \sum_{t=0}^{T} \beta^t \left| \int_{\mathsf{X}} g_{n,\overline{\varphi}_n}(x) R_n^t(dx|\gamma) \right. \\
&\qquad \left. - \int_{\mathsf{X}} g_{\overline{\varphi}_n}(x) P_n^t(dx|\gamma) \right| + 2 \sum_{t=T+1}^{\infty} \beta^t \|g\| \Bigg).
\end{aligned}$$

Since the first term in the last expression converges to zero as $n \to \infty$ for any T by Lemma 6.6 and the second term in the last expression converges to zero as $T \to \infty$ by $\|g\| < \infty$, the proof is complete. □

The next theorem is the main result of this section. Before stating the theorem, we introduce the following notation. We denote by ($\mathbf{CP}^{\varepsilon}$), ($\mathbf{CP}_n^{\varepsilon}$), and ($\overline{\mathbf{CP}}_n^{\varepsilon}$) the constrained Markov decision problems defined in Section 6.3.1 when the constraint vector is given by $\boldsymbol{k} - \varepsilon \mathbf{1}$.

Theorem 6.8 *For any* $\kappa > 0$ *there exist* $\varepsilon > 0$ *and* $n \geq 1$ *such that if* $\overline{\varphi}_n$ *is an optimal policy for* ($\overline{\mathbf{CP}}_n^{\varepsilon}$) *obtained by extending an optimal policy* φ_n *for* ($\mathbf{CP}_n^{\varepsilon}$) *to* X, *then* $\overline{\varphi}_n \in \Phi$ *is feasible for* (**CP**) *and the true cost of* $\overline{\varphi}_n$ *is within* κ *of the value function of* (**CP**).

Proof We observe that one can recover all the results derived in Section 6.3.1 if the constraint vector $\boldsymbol{k}$ is replaced by $\boldsymbol{k} - \varepsilon \mathbf{1}$, where $\varepsilon > 0$ satisfies

$$\varepsilon < \min_{l=1,\ldots,q} \alpha_l'. \tag{6.15}$$

Here, $\boldsymbol{\alpha}' = (\alpha_1', \ldots, \alpha_q')$ is the vector in Assumption 6.2-(b). Similar to the set $\mathscr{C}_n$ and the function $\mathscr{V}_n$ in the proof of Proposition 6.4, define the set $\mathscr{C} \subset \mathbb{R}^q$ and the function $\mathscr{V} : \mathscr{C} \to \mathbb{R}$ as

$$\mathscr{C} = \bigcup_{\zeta \in \mathscr{P}} \left\{ \boldsymbol{m} \in \mathbb{R}^q : \langle \zeta, c \rangle + \boldsymbol{\alpha} = \boldsymbol{m} \text{ for some } \boldsymbol{\alpha} \in \mathbb{R}^q_+ \right\},$$

$$\mathscr{V}(\boldsymbol{m}) = \min \left\{ \langle \zeta, c \rangle : \zeta \in \mathscr{P} \text{ and } \langle \zeta, \boldsymbol{d} \rangle + \boldsymbol{\alpha} = \boldsymbol{m},\ \boldsymbol{\alpha} \in \mathbb{R}^q_+ \right\}.$$

Hence, $\mathscr{C}$ is a convex subset of $\mathbb{R}^q$ and $\mathscr{V}$ is a convex function. We also have $\mathscr{V}(\boldsymbol{k} - \varepsilon \mathbf{1}) = \min(\mathbf{CP}^{\varepsilon})$ for any $\varepsilon \geq 0$. Since $\boldsymbol{k} \in \operatorname{int}\mathscr{C}$, the function $\mathscr{V}$, being convex, is continuous at $\boldsymbol{k}$. This implies the existence of a sequence $\{\varepsilon_k\}_{k \geq 1}$ of positive real numbers such that (i) ε_k satisfies (6.15) for all k, (ii) $\lim_{k \to \infty} \varepsilon_k = 0$, and therefore, $\lim_{k \to \infty} \mathscr{V}(\boldsymbol{k} - \varepsilon_k \mathbf{1}) = \mathscr{V}(\boldsymbol{k})$.

Given any $\kappa > 0$, we choose $k \geq 1$ sufficiently large such that

$$\left| \mathscr{V}(\boldsymbol{k}) - \mathscr{V}(\boldsymbol{k} - \varepsilon_k \mathbf{1}) \right| < \frac{\kappa}{3}. \tag{6.16}$$

Then, for $\boldsymbol{k} - \varepsilon_k \mathbf{1}$, we choose n sufficiently large such that

$$\left| \mathscr{V}(\boldsymbol{k} - \varepsilon_k \mathbf{1}) - \min(\mathbf{CP}_n^{\varepsilon_k}) \right| < \frac{\kappa}{3} \tag{6.17}$$

$$\left| J^c(\overline{\varphi}_n, \gamma) - J_n^c(\overline{\varphi}_n, \gamma) \right| = \left| J(\overline{\varphi}_n, \gamma) - \min(\mathbf{CP}_n^{\varepsilon_k}) \right| < \frac{\kappa}{3} \tag{6.18}$$

$$\left| J^{d_l}(\overline{\varphi}_n, \gamma) - J_n^{d_l}(\overline{\varphi}_n, \gamma) \right| < \varepsilon_k \text{ for } l = 1, \ldots, q, \tag{6.19}$$

where $\overline{\varphi}_n$ is the optimal policy for $(\overline{\mathbf{CP}}_n^{\varepsilon_k})$ obtained by extending the optimal policy φ_n of $(\mathbf{CP}_n^{\varepsilon_k})$ to X, i.e., $\overline{\varphi}_n(\cdot|x) = \varphi(\cdot|Q_n(x))$. Here, (6.17) follows from Theorem 6.1; (6.18), and (6.19) follow from Proposition 6.7. We observe that (6.19) implies that $\overline{\varphi}_n$ is feasible for (**CP**), and furthermore, by (6.16), (6.17), and (6.18), the true cost of $\overline{\varphi}_n$ is within κ of the value function of (**CP**), i.e.,

$$\left| J(\overline{\varphi}_n, \gamma) - \min(\mathbf{CP}) \right| < \kappa.$$

This completes the proof. □

6.4 Approximation of Average Cost Problems

In this section we obtain approximation results, analogous to Theorems 6.1 and 6.8, for the average cost criterion. To achieve this, we impose Assumption 4.2. Furthermore, we suppose that there exists a policy $\pi \in \Pi$ such that $V_l(\pi, \gamma) < k_l$ for $l = 1, \ldots, q$.

Note that any randomized stationary policy φ defines a stochastic kernel

$$p(\,\cdot\,|x,\varphi) := \int_{\mathsf{A}} p(\,\cdot\,|x,a)\varphi(da|x)$$

on X given X which is the transition probability of the Markov chain $\{X_t\}_{t=0}^{\infty}$ (state process) induced by φ. For any $t \geq 1$, we let $p^t(\,\cdot\,|x,\varphi)$ denote the t-step transition probability of this Markov chain given the initial point x. Recall from Theorem 4.12 that under above assumptions, the stochastic kernel $p(\,\cdot\,|x,\varphi)$ is positive Harris recurrent with unique invariant probability measure μ_φ, and we have

$$V(\varphi,\gamma) = \langle \mu_\varphi, c_\varphi \rangle,$$

and

$$V_l(\varphi,\gamma) = \langle \mu_\varphi, d_{l,\varphi} \rangle, \; l = 1,\ldots,q.$$

Furthermore, there exist positive real numbers R and $\kappa < 1$ such that for every $x \in \mathsf{X}$

$$\sup_{\varphi \in \Phi} \| p^t(\,\cdot\,|x,\varphi) - \mu_\varphi \|_{TV} \leq R\kappa^t,$$

where R and κ continuously depend on α.

The preceding results and [68, Lemma 5.7.10] imply that **(CP)** is equivalent to the following optimization problem, which is also denoted by **(CP)**:

$$\textbf{(CP)} \text{ minimize } \langle \mu_\varphi, c_\varphi \rangle$$
$$\text{subject to } \langle \mu_\varphi, d_{l,\varphi} \rangle \leq k_l \; (l = 1,\ldots,q), \; \varphi \in \Phi.$$

Furthermore, recall the definitions

$$\phi_n(x,a) := \int \phi(y,a)\nu_{n,i_n(x)}(dy),$$
$$\lambda_n := Q_n * \lambda.$$

Then $\overline{\text{MDP}}_n$ satisfies Assumption 4.2-(d),(e) when ϕ is replaced by ϕ_n, and Assumption 4.2-(d),(e) is true for MDP_n when ϕ and λ are replaced by the restriction of ϕ_n to X_n and λ_n, respectively. Hence, the above results holds (with the same R and κ) for $\overline{\text{MDP}}_n$ and MDP_n for all n. We denote by $\bar{\mu}_\varphi^n$ and μ_φ^n the invariant probability measures of $\overline{\text{MDP}}_n$ and MDP_n corresponding to the policy φ, respectively. Therefore, the average cost constrained problems for MDP_n and $\overline{\text{MDP}}_n$ are equivalent to the following optimization problems, respectively:

$$(\mathbf{CP}_n) \text{ minimize } \langle \mu_{\varphi}^n, c_{n,\varphi} \rangle$$
$$\text{subject to } \langle \mu_{\varphi}^n, d_{l,n,\varphi} \rangle \leq k_l \ (l = 1, \ldots, q), \ \varphi \in \Phi_n.$$

$$(\overline{\mathbf{CP}}_n) \text{ minimize } \langle \bar{\mu}_{\varphi}^n, b_{n,\varphi} \rangle$$
$$\text{subject to } \langle \bar{\mu}_{\varphi}^n, r_{l,n,\varphi} \rangle \leq k_l \ (l = 1, \ldots, q), \ \varphi \in \Phi.$$

The following lemma can be proved similar to Lemma 4.15.

Lemma 6.9 *For all $t \geq 1$, we have*

$$\lim_{n \to \infty} \sup_{(y,\varphi) \in \mathsf{X} \times \Phi} \left\| p^t(\,\cdot\,|y, \varphi) - q_n^t(\,\cdot\,|y, \varphi) \right\|_{TV} = 0.$$

Using Lemma 6.9, one can also prove, similar to Lemmas 4.16 and 4.18, that

$$\lim_{n \to \infty} \sup_{\varphi \in \Phi} \left| \langle \mu_{\varphi}, c_{\varphi} \rangle - \langle \bar{\mu}_{\varphi}^n, b_{n,\varphi} \rangle \right| = 0, \tag{6.20}$$

$$\lim_{n \to \infty} \sup_{\varphi \in \Phi} \left| \langle \mu_{\varphi}, d_{l,\varphi} \rangle - \langle \bar{\mu}_{\varphi}^n, r_{l,n,\varphi} \rangle \right| = 0, \ l = 1, \ldots, q. \tag{6.21}$$

Therefore, (6.21) implies that there exists $n_f \in \mathbb{N}$ such that for $n \geq n_f$, the problem $(\overline{\mathbf{CP}}_n)$ is consistent; that is, there exists a policy $\varphi \in \Phi$ which satisfies the constraints in $(\overline{\mathbf{CP}}_n)$. By [73, Theorem 4.5], we can also conclude that $(\mathbf{CP}_n)$ is also consistent. Then, by [73, Theorem 3.2], we have the following result.

Theorem 6.10 *Each of (**CP**), (**CP**$_n$), and ($\overline{\mathbf{CP}}_n$) is solvable; that is, there exists an optimal policy for each problem.*

In the remainder of this section, it is assumed that $n \geq n_f$. Analogous to Lemma 6.2, the following result states that MDP_n and $\overline{\text{MDP}}_n$ are essentially equivalent for the average cost.

Lemma 6.11 *We have*

$$\min(\boldsymbol{CP}_n) = \min(\overline{\boldsymbol{CP}}_n),$$

*and if the randomized stationary policy $\varphi^* \in \Phi_n$ is optimal for (**CP**$_n$), then its extension $\overline{\varphi}^*$ to X is also optimal for ($\overline{\mathbf{CP}}_n$) with the same cost function, where $\overline{\varphi}^*$ is given by $\overline{\varphi}^*(\,\cdot\,|x) = \varphi^*(\,\cdot\,|Q_n(x))$.*

Proof Analogous to the proof of Lemma 6.2, we first prove that any policy $\overline{\varphi} \in \Phi$, which is an extension (to X) of a feasible policy $\varphi \in \Phi_n$ for $(\mathbf{CP}_n)$, is also feasible for $(\overline{\mathbf{CP}}_n)$.

Fix any $\varphi \in \Phi_n$ feasible for $(\mathbf{CP}_n)$ and extend φ to X by letting $\overline{\varphi}(\cdot|x) = \varphi(\cdot|Q_n(x))$. It can be proved that

$$\mu_{\varphi}^{n}(\cdot) = Q_n * \bar{\mu}_{\overline{\varphi}}^{n}.$$

Then we have

$$\langle \bar{\mu}_{\overline{\varphi}}^{n}, b_{n,\overline{\varphi}} \rangle = \langle \mu_{\varphi}^{n}, c_{n,\varphi} \rangle$$

$$\langle \bar{\mu}_{\overline{\varphi}}^{n}, r_{l,n,\overline{\varphi}} \rangle = \langle \mu_{\varphi}^{n}, d_{l,n,\varphi} \rangle, \; l = 1, \ldots, q,$$

which proves the first result. Therefore, we have $\min(\mathbf{CP}_n) \geq \min(\overline{\mathbf{CP}}_n)$.

To complete the proof it is sufficient to prove $\min(\mathbf{CP}_n) \leq \min(\overline{\mathbf{CP}}_n)$. This can be done by formulating both $(\mathbf{CP}_n)$ and $(\overline{\mathbf{CP}}_n)$ as linear programs [73] and then proving that the dual linear program $(\mathbf{CP}_n^*)$ of $(\mathbf{CP}_n)$ has a smaller value function than the dual linear program $(\overline{\mathbf{CP}}_n^*)$ of $(\overline{\mathbf{CP}}_n)$. Indeed, the latter can be proved by showing that the feasible points of $(\mathbf{CP}_n^*)$ can be embedded into the feasible points of $(\overline{\mathbf{CP}}_n^*)$ with the same cost function. As there is no duality gap for both problems [73, Theorem 4.4], we have $\min(\mathbf{CP}_n) = \sup(\mathbf{CP}_n^*) \leq \sup(\overline{\mathbf{CP}}_n^*) = \min(\overline{\mathbf{CP}}_n)$. Since the linear programming formulation of the constrained average-cost problems will not be used in the sequel, we omit the details and refer the reader to [73] to fill the technical gaps in the above discussion. □

Making use of Lemma 6.11, in the remainder of this section we consider $\overline{\text{MDP}}_n$ in place of MDP_n. We define

$$\Delta(\boldsymbol{m}) := \{\varphi \in \Phi : \langle \mu_{\varphi}, d_{l,\varphi} \rangle \leq m_l, \; l = 1, \ldots, q\}$$

$$\Delta_n(\boldsymbol{m}) := \{\varphi \in \Phi : \langle \bar{\mu}_{\varphi}^{n}, r_{l,n,\varphi} \rangle \leq m_l, \; l = 1, \ldots, q\}.$$

Then, we let $\mathscr{C} := \{\boldsymbol{m} \in \mathbb{R}^q : \Delta(\boldsymbol{m}) \neq \emptyset\}$ and $\mathscr{C}_n := \{\boldsymbol{m} \in \mathbb{R}^q : \Delta_n(\boldsymbol{m}) \neq \emptyset\}$. It can be proved that both $\mathscr{C}$ and $\mathscr{C}_n$ are convex subsets of $\mathbb{R}^q$. Let us also define functions $\mathscr{W}$ and $\mathscr{W}_n$ over $\mathscr{C}$ and $\mathscr{C}_n$, respectively, as follows:

$$\mathscr{W}(\boldsymbol{m}) := \min\{\langle \mu_{\varphi}, c_{\varphi} \rangle : \varphi \in \Delta(\boldsymbol{m})\}$$

$$\mathscr{W}_n(\boldsymbol{m}) := \min\{\langle \bar{\mu}_{\varphi}^{n}, b_{n,\varphi} \rangle : \varphi \in \Delta_n(\boldsymbol{m})\}.$$

It can also be proved that both $\mathscr{W}$ and $\mathscr{W}_n$ are convex functions. Note that $\min(\mathbf{CP}) = \mathscr{W}(\boldsymbol{k})$ and $\min(\overline{\mathbf{CP}}_n) = \mathscr{W}_n(\boldsymbol{k})$. Furthermore, by assumption we have $\boldsymbol{k} \in \operatorname{int} \mathscr{C}$. Since, $n \geq n_f$, we also have $\boldsymbol{k} \in \operatorname{int} \mathscr{C}_n$. Therefore, functions $\mathscr{W}_n$ and $\mathscr{W}$, being convex, are continuous at $\boldsymbol{k}$.

The following theorem is analogous to Theorem 6.1 and states that the value function of $(\mathbf{CP}_n)$ (or equivalently, the value function of $(\overline{\mathbf{CP}}_n)$) converges to the value function of $(\mathbf{CP})$.

Theorem 6.12 *We have*

$$\lim_{n\to\infty} \left| \min(\boldsymbol{CP}_n) - \min(\boldsymbol{CP}) \right|.$$

Proof The result follows from (6.20) and (6.21), and the fact that $\mathscr{W}_n$ and $\mathscr{W}$ are continuous at $\boldsymbol{k}$. □

The following theorem is analogous to Theorem 6.8 and is the main result of this section. Similar to the notation introduced preceding to Theorem 6.8, we denote by $(\mathbf{CP}^{\varepsilon})$, $(\mathbf{CP}_n^{\varepsilon})$, and $(\overline{\mathbf{CP}}_n^{\varepsilon})$ the average cost constrained Markov decision processes defined above when the constraint vector is given by $\boldsymbol{k} - \varepsilon\mathbf{1}$.

Theorem 6.13 *For any given $\kappa > 0$, there exist $\varepsilon > 0$ and $n \geq 1$ such that if $\overline{\varphi}_n$ is an optimal policy for $(\overline{\boldsymbol{CP}}_n^{\varepsilon})$ obtained by extending an optimal policy φ_n for $(\boldsymbol{CP}_n^{\varepsilon})$ to X, then $\overline{\varphi}_n \in \Phi$ is feasible for $(\boldsymbol{CP})$ and the true cost of $\overline{\varphi}_n$ is within κ of the value function of $(\boldsymbol{CP})$.*

Proof The result follows from (6.20) and (6.21), the fact that $\mathscr{W}_n$ and $\mathscr{W}$ are continuous at $\boldsymbol{k}$, and Theorem 6.12. It can be done similar to the proof of Theorem 6.8, and so we omit the details. □

6.5 Concluding Remarks

In this chapter, the approximation of discrete-time constrained Markov decision processes with compact state spaces is considered. By formulating the constrained discounted problem as linear program, we first showed that the optimal value function of the reduced model converges to the optimal value function of the original model. Then, under the total variation continuity of the transition probability, we developed a method which results in approximately optimal policies. Under drift and minorization conditions on the transition probability, we derived analogous approximation results for the average cost. However, in this case, we used a direct method, instead of linear programming formulation, to establish the results.

We refer the reader to [41, 43, 45] for a further reading on finite model approximation problem for constrained MDPs. In [41, 43], the authors consider finite linear programming approximation of constrained Markov decision process with discounted cost. They establish a sequence of approximate models using, respectively, quantized and empirical distributions of a probability measure μ with respect to which the transition probability of the MDP is absolutely continuous. Under Lipschitz type continuity conditions on the system components, they establish the convergence of the optimal value function of the approximate model to the optimal value function of the constrained MDP. They also provide a rate of convergence result. In [45], the authors consider finite linear programming approximations of constrained Markov decision process with compact state-action pair for both the discounted cost and average cost criteria. The authors provide explicit error bounds for the approximation of value function under Lipschitz type regularity assumption on the transition probability and the one-stage cost function. However, these approaches are not constructive, since they do not yield approximately optimal policies for the original problem through finite models.

Some of the results presented in this chapter are based on [117].

Part II
Finite Model Approximations in Decentralized Stochastic Control

Chapter 7
Prelude to Part II

In Part II, we focus on decentralized stochastic control problems and their applications. In Chapter 8, we present our results on the finite model approximation of multi-agent stochastic control problems (team decision problems). We show that optimal strategies obtained from finite models approximate the optimal cost with arbitrary precision under mild technical assumptions. In particular, we show that quantized team policies are asymptotically optimal. In Chapter 9, the results are applied to Witsenhausen's counterexample and the Gaussian relay channel problem.

In this chapter, we provide some preliminaries on decentralized stochastic control.

7.1 Witsenhausen's Intrinsic Model

This section will introduce the team decision model as laid out by Witsenhausen, called *the Intrinsic Model* [142]. We refer to [148] for a more comprehensive overview and further characterizations and classifications of information structures. In this model, any action applied at any given time is regarded as being applied by an individual decision maker (or agent), who acts only once. For instance, in the intrinsic model for an MDP, the action A_t is assumed to be applied by a distinct individual decision maker. One advantage of this model for the multi-agent setup, in addition to its generality, is that the definitions regarding information structures can be compactly described.

Suppose that in a decentralized system there is a predefined order in which the agents act. Such a system is called a *sequential system* (for nonsequential teams, we refer the reader to [4, 5] and [125], in addition to [140]). In the following, all spaces are assumed to be Borel spaces endowed with Borel σ-algebras. In the context of a sequential system, the *Intrinsic Model* has the following components:

N. Saldi et al., *Finite Approximations in Discrete-Time Stochastic Control*,
Systems & Control: Foundations & Applications,
https://doi.org/10.1007/978-3-319-79033-6_7

- A collection of *measurable spaces*

$$\mathscr{I} := \{(\mathsf{X}, \mathscr{B}(\mathsf{X})), (\mathsf{U}^i, \mathscr{B}(\mathsf{U}^i)), (\mathsf{Y}^i, \mathscr{B}(\mathsf{Y}^i)), i \in \mathscr{N}\},$$

  specifying the system's distinguishable events, and the action and measurement spaces. Here $|\mathscr{N}| := N$ is the number of actions taken, and each of these actions is supposed to be taken by an individual agent (hence, an agent with perfect recall can also be regarded as a separate decision maker every time it acts). Here X is a space on which an underlying probability can be defined. The space U^i denotes the space from which the action U^i of Agent i is selected and Y^i denotes the measurement (or observation) space of Agent i.
- A *measurement constraint* which establishes the connection between the observation variables and the system's distinguishable events. The Y^i-valued observation variables are given by $Y^i \sim W^i(\cdot|X, \boldsymbol{U}^{i-1})$, where $\boldsymbol{U}^{i-1} = (U^1, \ldots, U^{i-1})$, W^i is a stochastic kernel on Y^i given $\mathsf{X} \times \prod_{j=1}^{i-1} \mathsf{U}^j$, and U^k denotes the action of Agent k.
- A *design constraint*, which restricts the set of admissible N-tuple control strategies $\underline{\gamma} = (\gamma^1, \gamma^2, \ldots, \gamma^N)$, also called *policies*, to the set of all measurable functions, so that $U^i = \gamma^i(Y^i)$, where γ^i is a measurable function. Let Γ^i denote the set of all admissible policies for Agent i and let $\boldsymbol{\Gamma} = \prod_k \Gamma^k$.
- A *probability measure* $\mathbb{P}$ on $(\mathsf{X}, \mathscr{B}(\mathsf{X}))$ which describes the measures on the random events in the model.

We note that the intrinsic model of Witsenhausen gives a set-theoretic characterization of information fields; however, for Borel spaces, the model above is equivalent to the intrinsic model for sequential team problems.

Under this intrinsic model, a sequential team problem is *dynamic* if the information Y^i available to at least one agent i is affected by the action of at least one other agent $k \neq i$; that is, there exist $i \in \mathscr{N}$ and $\boldsymbol{u}^{i-1}, \tilde{\boldsymbol{u}}^{i-1} \in \prod_{j=1}^{i-1} \mathsf{U}^j$ such that $W^i(\cdot|X, \boldsymbol{u}^{i-1}) \neq W^i(\cdot|X, \tilde{\boldsymbol{u}}^{i-1})$. A decentralized problem is *static* if the information available at every decision maker is only affected by state of the nature; that is, for all $i \in \mathscr{N}$, we have $W^i(\cdot|X, \boldsymbol{U}^{i-1}) = W^i(\cdot|X)$.

Information structures (ISs) can also be classified as classical, quasi-classical, and nonclassical. An IS is *classical* if, for all $i \in \mathscr{N}$, Y^i contains all of the information available to Agent k for all $k < i$; that is, for all $k < i$, there exists a measurable function $g_{i,k} : \mathsf{Y}^i \to \mathsf{Y}^k$ such that $Y^k = g_{i,k}(Y^i)$ for all $\underline{\gamma} \in \boldsymbol{\Gamma}$ and all realizations of X. An IS is *quasi-classical* or *partially nested*, if whenever U^k (for some $k < i$) affects Y^i, then Agent i has access to Y^k; that is, there exists a measurable function $g_{i,k} : \mathsf{Y}^i \to \mathsf{Y}^k$ such that $g_{i,k}(Y^i) = Y^k$ for all $\underline{\gamma} \in \boldsymbol{\Gamma}$ and all realizations of X. For such team problems, one talks about *precedence relationships* among agents: Agent k is *precedent* to Agent i (or Agent k *communicates* to Agent i), if the former agent's actions affect the information of the latter, in which case (to be partially nested) Agent i has to have the information based on which the action-generating policy of Agent k was constructed. An IS which is not partially nested is *nonclassical*.

For any $\underline{\gamma} = (\gamma^1, \cdots, \gamma^N)$, we let the cost of the team problem be defined by

$$J(\underline{\gamma}) := \mathbb{E}\big[c(X, \boldsymbol{Y}, \boldsymbol{U})\big],$$

for some measurable cost function $c : \mathsf{X} \times \prod_i \mathsf{Y}^i \times \prod_i \mathsf{U}^i \to \mathbb{R}_+$, where $\boldsymbol{U} := (U^1, \ldots, U^N) = \underline{\gamma}(\boldsymbol{Y})$ and $\boldsymbol{Y} := (Y^1, \ldots, Y^N)$. For a given stochastic team problem, a policy (strategy) $\underline{\gamma}^* := (\gamma^{1*}, \ldots, \gamma^{N*}) \in \boldsymbol{\Gamma}$ is an *optimal team decision rule* if

$$J(\underline{\gamma}^*) = \inf_{\underline{\gamma} \in \boldsymbol{\Gamma}} J(\underline{\gamma}) =: J^*.$$

The cost level J^* achieved by this strategy is the *optimal team cost*. For a given stochastic team problem, a policy $\underline{\gamma}^* := (\gamma^{1*}, \ldots, \gamma^{N*})$ constitutes a *Nash equilibrium* (a *person-by-person optimal* solution) if, for all $\beta \in \Gamma^i$ and all $i \in N$, the following inequalities hold:

$$J^* := J(\underline{\gamma}^*) \leq J(\underline{\gamma}^{-i*}, \beta),$$

where we have adopted the notation

$$(\underline{\gamma}^{-i*}, \beta) := (\gamma^{1*}, \ldots, \gamma^{i-1*}, \beta, \gamma^{i+1*}, \ldots, \gamma^{N*}).$$

In team decision theory, the terms *policy*, *measurement*, and *agent* are used synonymously with *strategy*, *observation*, and *decision maker*, respectively.

For a class of teams which are *convex*, under mild technical conditions, person-by-person-optimality implies global optimality (see [37, 80, 105, 148] and one can reduce the search space to a smaller parametric class of policies. For a comprehensive review, see [148]).

However, in general establishing the existence and structure of optimal policies is a challenging problem. Existence of optimal policies for static and a class of sequential dynamic teams has been studied recently in [60]; see also [138, 144] for the specific context of Witsenhausen's counterexample.

7.2 Static Reduction of Sequential Dynamic Teams

In this section, we review the equivalence between sequential dynamic teams and their static reduction (this is called *the equivalent model* [143]). Consider a dynamic team setting according to the intrinsic model where there are N decision epochs, and Agent i observes $Y^i \sim W^i(\cdot \,|X, \boldsymbol{U}^{i-1})$, and the decisions are generated as

$U^i = \gamma^i(Y^i)$. The resulting cost under a given team policy $\underline{\gamma}$ is

$$J(\underline{\gamma}) = \mathbb{E}[c(X, \boldsymbol{Y}, \boldsymbol{U})].$$

This dynamic team can be converted to a static team provided that the following absolute continuity condition holds.

Assumption 7.1 *For every $i = 1, \ldots, N$, there exists a function $f_i : \mathsf{X} \times \mathsf{U}_1^{i-1} \times \mathsf{Y}^i \to [0, \infty)$, where $\mathsf{U}_1^{i-1} := \prod_{j=1}^{i-1} \mathsf{U}^i$, and a probability measure Q_i on Y^i such that for all $S \in \mathscr{B}(\mathsf{Y}^i)$ we have*

$$W^i(S|x, \boldsymbol{u}^{i-1}) = \int_S f_i(x, \boldsymbol{u}^{i-1}, y^i) Q_i(dy^i).$$

Therefore, for a fixed choice of $\underline{\gamma}$, the joint distribution of $(X, \boldsymbol{Y})$ is given by

$$\mathbb{P}(dx, d\boldsymbol{y}) = \mathbb{P}(dx) \prod_{i=1}^{N} f_i(x, \boldsymbol{u}^{i-1}, y^i) Q_i(dy^i),$$

where $\mathbf{u}^{i-1} = \big(\gamma^1(y^1), \ldots, \gamma^{i-1}(y^{i-1})\big)$. The cost function $J(\underline{\gamma})$ can then be written as

$$\begin{aligned} J(\underline{\gamma}) &= \int_{\mathsf{X}\times\mathsf{Y}} c(x, \boldsymbol{y}, \boldsymbol{u}) \mathbb{P}(dx) \prod_{i=1}^{N} f_i(x, \boldsymbol{u}^{i-1}, y^i) Q_i(dy^i) \\ &= \int_{\mathsf{X}\times\mathsf{Y}} \tilde{c}(x, \boldsymbol{y}, \boldsymbol{u}) \widetilde{\mathbb{P}}(dx, d\boldsymbol{y}), \end{aligned}$$

where $\tilde{c}(x, \boldsymbol{y}, \boldsymbol{u}) := c(x, \boldsymbol{y}, \boldsymbol{u}) \prod_{i=1}^{N} f_i(x, \boldsymbol{u}^{i-1}, y^i)$ and $\widetilde{\mathbb{P}}(dx, d\boldsymbol{y}) := \mathbb{P}(dx) \prod_{i=1}^{N} Q_i(dy^i)$. The observations now can be regarded as independent, and by incorporating the f_i terms into c, we can obtain an equivalent *static team* problem. Hence, the essential step is to appropriately adjust the probability space and the cost function.

Remark 7.1 Note that in the static reduction method some nice properties (such as continuity and boundedness) of the cost function c of the original dynamic team problem can be lost if the f_i functions in Assumption 7.1 are not well behaved. However, the observation channels between X and the Y^i are quite well behaved for most of the practical models (i.e., additive Gaussian channel) admitting static reduction. This usually implies that most nice properties of the cost function are preserved for most of the practical models.

As Witsenhausen notes in [143], above static reduction always holds when the measurement variables take values from countable set since any probability measure on that countable set has a density with respect to weighted counting measure $Q_i(\cdot) := \sum_{k\geq 1} 2^{-k}\delta_{y_k}(\cdot)$, where $\mathsf{Y}^i = \{y_k, k \in \mathbb{N}\}$. On the other hand, for continuous spaces, the absolutely continuity condition (i.e., Assumption 7.1) does not hold for some practically relevant systems. For instance, suppose that $Y^2 = U^1$; that is, Agent 2 can only observe the action of Agent 1. Then, the absolute continuity condition required for Witsenhausen's static reduction fails. We also note that a continuous-time generalization for Witsenhausen's static reduction similar to Girsanov's method has been presented in [31].

Another important dynamic information structure, for which we can establish a static reduction, is the *partially nested* information structure. For partially nested (or quasi-classical) information structures, a static reduction was studied by Ho and Chu in the context of LQG systems [76] and for a class of nonlinear systems satisfying restrictive invertibility properties [77]. For such dynamic teams with a static reduction, the cost function does not change as a result of the static reduction, unlike in the static reduction introduced previously. A general discussion on static reduction and its relation with convexity properties is available in [150].

Chapter 8
Finite Model Approximations in Decentralized Stochastic Control

8.1 Introduction

In this chapter, we study the approximation of static and dynamic team problems using finite models which are obtained through the uniform discretization, on a finite grid, of the observation and action spaces of agents. In particular, we are interested in the asymptotic optimality of quantized policies.

8.1.1 Auxiliary Results

Here we state some well-known results in measure theory and functional analysis that will be frequently used in the chapter. The first result is Lusin's theorem which roughly states that any measurable function is *almost* continuous.

Theorem 8.1 (Lusin's Theorem [39, Theorem 7.5.2]) *Let* E_1 *and* E_2 *be two Borel spaces and let* μ *be a probability measure on* E_1. *Assume* $f : \mathsf{E}_1 \to \mathsf{E}_2$ *is measurable. Then, for any* $\varepsilon > 0$, *there is a closed set* $F \subset \mathsf{E}_1$ *such that* $\mu(\mathsf{E}_1 \setminus F) < \varepsilon$ *and the restriction of* f *to* F *is continuous.*

The second theorem is the Dugundji extension theorem which is a generalization of the Tietze extension theorem [39].

Theorem 8.2 (Dugundji Extension Theorem [56, Theorem 7.4]) *Let* E_1 *be a Borel space and let* F *be a closed subset of* E_1. *Let* E_2 *be a convex subset of some locally convex vector space. Then any continuous* $f : F \to \mathsf{E}_2$ *has a continuous extension on* E_1.

The next theorem originally states that the closed convex hull of a compact subset in a locally convex vector space E is compact if the vector space is completely

N. Saldi et al., *Finite Approximations in Discrete-Time Stochastic Control*, Systems & Control: Foundations & Applications,
https://doi.org/10.1007/978-3-319-79033-6_8

metrizable (i.e., there exists a metric d on E such that (E, d) is a complete metric space). Since a closed subset of a compact set is compact, we can state the theorem in the following form.

Theorem 8.3 ([2, Theorem 5.35]) *In a completely metrizable locally convex vector space* E*, the closed convex hull of a compact set is compact. The same statement also holds when* E *is replaced with any of its closed and convex subsets.*

8.2 Approximation of Static Team Problems

In this section, we consider the finite observation approximation of static team problems, where these approximations are obtained through the uniform discretization, on a finite grid, of the observation spaces. We establish that finite observation models asymptotically represent the true models in the sense that the solutions obtained by solving such finite models lead to cost values that converge to the optimal cost of the original model. Since the problem is static, the observation channel $W^i(\,\cdot\,|X)$ of Agent i, $i = 1, \ldots, N$, only depends on the variable X; that is,

$$Y^i \sim W^i(\,\cdot\,|X), \ \text{for } i = 1, \ldots, N.$$

Recall that the team cost function c is a nonnegative function of the state, observations, and actions; that is, $c : \mathsf{X} \times \mathsf{Y} \times \mathsf{U} \to [0, \infty)$, where $\mathsf{Y} := \prod_{i=1}^N \mathsf{Y}^i$ and $\mathsf{U} := \prod_{i=1}^N \mathsf{U}^i$. For Agent i, the set of strategies Γ^i is given by

$$\Gamma^i := \big\{\gamma^i : \mathsf{Y}^i \to \mathsf{U}^i, \gamma^i \text{ is measurable}\big\}$$

and $\boldsymbol{\Gamma} = \prod_{i=1}^N \Gamma^i$. Then, the cost of the team, $J : \boldsymbol{\Gamma} \to [0, \infty)$, is given by

$$J(\underline{\gamma}) = \int_{\mathsf{X}\times\mathsf{Y}} c(x, \boldsymbol{y}, \boldsymbol{u})\mathbb{P}(dx, d\boldsymbol{y}),$$

where $\boldsymbol{u} = \underline{\gamma}(\boldsymbol{y})$. Here, with an abuse of notation, $\mathbb{P}(dx, d\boldsymbol{y}) := \mathbb{P}(dx)\prod_{i=1}^N W^i(dy^i|x)$ denotes the joint distribution of the state and observations. Therefore, we have

$$J^* = \inf_{\underline{\gamma}\in\boldsymbol{\Gamma}} J(\underline{\gamma}).$$

We first consider the case where the observation spaces are compact and the cost function is bounded. Then, teams with non-compact observation spaces and unbounded cost function will be studied.

8.2.1 Approximation of Static Teams with Compact Observation Spaces and Bounded Cost

We impose the following assumptions on the components of the model.

Assumption 8.1

(a) *The cost function c is bounded. In addition, it is continuous in $(\boldsymbol{u}, \boldsymbol{y})$ for any fixed x.*
(b) *For each i, U^i is a convex subset of a locally convex vector space.*
(c) *For each i, Y^i is compact.*

We first prove that the minimum cost achievable by continuous strategies is equal to the optimal cost J^*. To this end, we define $\Gamma_c^i := \{\gamma^i \in \Gamma^i : \gamma^i \text{ is continuous}\}$ and $\boldsymbol{\Gamma}_c := \prod_{i=1}^N \Gamma_c^i$ for all $i = 1, \ldots, N$.

Proposition 8.4 *We have*

$$\inf_{\underline{\gamma} \in \boldsymbol{\Gamma}_c} J(\underline{\gamma}) = J^*.$$

Proof Let $\underline{\gamma} \in \boldsymbol{\Gamma}$ be arbitrary. We prove that there exists a sequence $\{\underline{\gamma}_k\}_{k \geq 1} \in \boldsymbol{\Gamma}_c$ such that $J(\underline{\gamma}_k) \to J(\underline{\gamma})$ as $k \to \infty$, which implies the proposition. Let μ^i denote the distribution of Y^i.

For each $k \geq 1$, by Lusin's theorem, there are closed sets $F_{k,i} \subset \mathsf{Y}^i$ such that $\mu^i(\mathsf{Y}^i \setminus F_{k,i}) < 1/k$ and the restriction of γ^i to $F_{k,i}$ is continuous. Define $\pi_k^i = \gamma^i|_{F_{k,i}}$ so that $\pi_k^i : F_{k,i} \to \mathsf{U}^i$ is continuous. By the Dugundji extension theorem, there exists a continuous extension $\gamma_k^i : \mathsf{Y}^i \to \mathsf{U}^i$ of π_k^i. Therefore, $\underline{\gamma}_k = (\gamma_k^1, \ldots, \gamma_k^N) \in \boldsymbol{\Gamma}_c$ and $F_k := \prod_{i=1}^N F_{k,i}$ satisfies

$$\begin{aligned} \left|J(\underline{\gamma}) - J(\underline{\gamma}_k)\right| &= \left| \int_{\mathsf{X} \times \mathsf{Y}} \left[c(x, \boldsymbol{y}, \underline{\gamma}) - c(x, \boldsymbol{y}, \underline{\gamma}_k)\right] \mathbb{P}(dx, d\boldsymbol{y}) \right| \\ &\leq \int_{\mathsf{X} \times (\mathsf{Y} \setminus F_k)} \left|c(x, \boldsymbol{y}, \underline{\gamma}) - c(x, \boldsymbol{y}, \underline{\gamma}_k)\right| \mathbb{P}(dx, d\boldsymbol{y}) \\ &\leq 2\|c\| \, \mathbb{P}\big(\mathsf{X} \times (\mathsf{Y} \setminus F_k)\big), \end{aligned}$$

where $\|c\|$ is the maximum absolute value that c takes. Since $\mathbb{P}\big(\mathsf{X} \times (\mathsf{Y} \setminus F_k)\big) \leq \sum_{i=1}^N \mu^i(\mathsf{Y}^i \setminus F_{k,i}) \leq N/k$, we have $\lim_{k \to \infty} J(\underline{\gamma}_k) = J(\underline{\gamma})$. This completes the proof. □

Let d_i denote the metric on Y^i. Since Y^i is compact, there exist finite sets $\mathsf{Y}_n^i := \{y_{i,1}, \ldots, y_{i,i_n}\} \subset \mathsf{Y}^i$ such that each Y_n^i is an $1/n$-net in Y^i; that is, for any $y \in \mathsf{Y}^i$, we have

$$\min_{z \in \mathsf{Y}_n^i} d_i(y, z) < \frac{1}{n}.$$

Define function q_n^i mapping Y^i to Y_n^i by

$$q_n^i(y) := \arg\min_{z \in \mathsf{Y}_n^i} d_i(y, z),$$

where ties are broken so that q_n^i is measurable. If $\mathsf{Y}^i = [-M, M]$ for some $M \in \mathbb{R}_+$, one can choose the finite set Y_n^i such that q_n^i becomes a uniform quantizer. For any $\gamma^i \in \Gamma^i$, we let $\gamma^{n,i}$ denote the strategy $\gamma^i \circ q_n^i$. Define

$$\Gamma_n^i := \Gamma^i \circ q_n^i$$

and let $\boldsymbol{\Gamma}_n := \prod_{i=1}^N \Gamma_n^i$. The following theorem states that an optimal (or almost optimal) policy can be approximated with arbitrarily small approximation error for the induced costs by policies in $\boldsymbol{\Gamma}_n$, if n is sufficiently large.

Theorem 8.5 *We have*

$$\lim_{n \to \infty} \inf_{\underline{\gamma} \in \boldsymbol{\Gamma}_n} J(\underline{\gamma}) = J^*.$$

Proof For any ε, let $\underline{\gamma}_\varepsilon = (\gamma_\varepsilon^1, \ldots, \gamma_\varepsilon^N) \in \boldsymbol{\Gamma}_c$ denote an ε-optimal continuous strategy. The existence of such a strategy follows from Proposition 8.4. Then we have

$$\begin{aligned} \inf_{\underline{\gamma} \in \boldsymbol{\Gamma}_n} J(\underline{\gamma}) - J^* &= \inf_{\underline{\gamma} \in \boldsymbol{\Gamma}_n} J(\underline{\gamma}) - \inf_{\underline{\gamma} \in \boldsymbol{\Gamma}_c} J(\underline{\gamma}) \text{ (by Proposition 8.4)} \\ &\leq J(\underline{\gamma}_{\varepsilon,n}) - \inf_{\underline{\gamma} \in \boldsymbol{\Gamma}_c} J(\underline{\gamma}) \\ &\leq \varepsilon + \big(J(\underline{\gamma}_{\varepsilon,n}) - J(\underline{\gamma}_\varepsilon)\big), \end{aligned}$$

where $\underline{\gamma}_{\varepsilon,n} = (\gamma_\varepsilon^{n,1}, \ldots, \gamma_\varepsilon^{n,N})$. Note that $c(x, \mathbf{y}, \underline{\gamma}_{\varepsilon,n}(\mathbf{y})) \to c(x, \mathbf{y}, \underline{\gamma}_\varepsilon(\mathbf{y}))$ as $n \to \infty$ for all $(x, \mathbf{y}) \in \mathsf{X} \times \mathsf{Y}$ since c is continuous in $\boldsymbol{u}$ and $\underline{\gamma}_\varepsilon \in \boldsymbol{\Gamma}_c$. Hence, by the dominated convergence theorem the second term in the last expression converges to zero as $n \to \infty$. Since ε is arbitrary, this completes the proof. □

For each n, define stochastic kernels $W_n^i(\cdot|x)$ on Y_n^i given X as follows:

$$W_n^i(\cdot|x) := \sum_{j=1}^{i_n} W(\mathcal{S}_{i,j}^n|x) \delta_{y_{i,j}}(\cdot),$$

where $\mathscr{S}_{i,j}^n := \{y \in \mathsf{Y}^i : q_n^i(y) = y_{i,j}\}$. Let $\Pi_n^i := \{\pi^i : \mathsf{Y}_n^i \to \mathsf{U}^i, \pi^i \text{ measurable}\}$ and $\boldsymbol{\Pi}_n := \prod_{i=1}^N \Pi_n^i$. Define $J_n : \boldsymbol{\Pi}_n \to [0, \infty)$ as

$$J_n(\underline{\pi}) := \int_{\mathsf{X}\times\mathsf{Y}_n} c(x, \mathbf{y}, \boldsymbol{u})\mathbb{P}_n(dx, d\mathbf{y}),$$

where $\underline{\pi} = (\pi^1, \ldots, \pi^N)$, $\boldsymbol{u} = \underline{\pi}(\mathbf{y})$, $\mathsf{Y}_n = \prod_{i=1}^N \mathsf{Y}_n^i$, and $\mathbb{P}_n(dx, d\mathbf{y}) = \mathbb{P}(dx)\prod_{i=1}^N W_n^i(dy^i|x)$.

Lemma 8.6 *Let $\{\underline{\pi}_n\}$ be a sequence of strategies such that $\underline{\pi}_n \in \boldsymbol{\Pi}_n$. For each n, define $\underline{\gamma}_n := \underline{\pi}_n \circ q_n$, where $q_n := (q_n^1, \ldots, q_n^N)$. Then, we have*

$$\lim_{n\to\infty} |J_n(\underline{\pi}_n) - J(\underline{\gamma}_n)| = 0.$$

Proof We have

$$\begin{aligned}
&|J_n(\underline{\pi}_n) - J(\underline{\gamma}_n)| \\
&= \left| \int_{\mathsf{X}\times\mathsf{Y}} c(x, q_n(\mathbf{y}), \underline{\gamma}_n)\, d\mathbb{P} - \int_{\mathsf{X}\times\mathsf{Y}} c(x, \mathbf{y}, \underline{\gamma}_n)\, d\mathbb{P} \right|
\end{aligned}$$

which converges to zero as $n \to \infty$ by dominated convergence theorem and the fact that c is bounded and continuous in $\mathbf{y}$. □

The following theorem, a consequence of Theorem 8.5, is the main result of this section. It states that to compute a near optimal strategy for the original team problem, it is sufficient to compute an optimal strategy (or almost optimal strategy if an optimal one does not exist) for the team problem with finite observation spaces described above.

Theorem 8.7 *For any $\varepsilon > 0$, there exists a sufficiently large n such that the optimal (or almost optimal) policy $\underline{\pi}^* \in \boldsymbol{\Pi}_n$ for the cost J_n is ε-optimal for the original team problem when $\underline{\pi}^* = (\pi^{1*}, \ldots, \pi^{N*})$ is extended to Y via $\gamma^i = \pi^{i*} \circ q_n^i$.*

Proof Fix any $\varepsilon > 0$. By Theorem 8.5, there exists a sequence of strategies $\{\underline{\gamma}_n\}$ such that $\underline{\gamma}_n \in \boldsymbol{\Gamma}_n$ and $\lim_{n\to\infty} J(\underline{\gamma}_n) = J^*$. Define $\underline{\pi}_n$ as the restriction of $\underline{\gamma}_n$ to the set Y_n. Then, we have

$$\begin{aligned}
J^* &= \lim_{n\to\infty} J(\underline{\gamma}_n) \\
&= \lim_{n\to\infty} J_n(\underline{\pi}_n) \text{ (by Lemma 8.6)} \\
&\geq \limsup_{n\to\infty} \inf_{\underline{\pi}\in\boldsymbol{\Pi}_n} J_n(\underline{\pi}).
\end{aligned}$$

For the reverse inequality, for each $n \geq 1$, let $\underline{\pi}_n \in \boldsymbol{\Pi}_n$ be such that $J_n(\underline{\pi}_n) < \inf_{\underline{\pi} \in \boldsymbol{\Pi}_n} J_n(\underline{\pi}) + 1/n$. Define $\underline{\gamma}_n := \underline{\pi}_n \circ q_n$. Then, we have

$$\begin{aligned}
\liminf_{n\to\infty} \inf_{\underline{\pi} \in \boldsymbol{\Pi}_n} J_n(\underline{\pi}) &= \liminf_{n\to\infty} J_n(\underline{\pi}_n) \\
&= \liminf_{n\to\infty} J(\underline{\gamma}_n) \text{ (by Lemma 8.6)} \\
&\geq J^*.
\end{aligned}$$

This completes the proof. □

8.2.2 *Approximation of Static Teams with Non-compact Observation Spaces and Unbounded Cost*

In this section, we consider the finite observation approximation of static team problems with non-compact observation spaces and unbounded cost function. We impose the following assumptions on the components of the model.

Assumption 8.2

(a) The cost function c is continuous in $\boldsymbol{u}$ and $\boldsymbol{y}$ for any fixed x. In addition, it is bounded on any compact subset of $\mathsf{X} \times \mathsf{Y} \times \mathsf{U}$.
(b) For each i, U^i is a closed and convex subset of a completely metrizable locally convex vector space.
(c) For each i, Y^i is locally compact.
(d) The function $w_G(x, \boldsymbol{y}) := \sup_{\boldsymbol{u} \in G} c(x, \boldsymbol{y}, \boldsymbol{u})$ is integrable with respect to $\mathbb{P}(dx, d\boldsymbol{y})$, for any compact subset G of U of the form $G = \prod_{i=1}^{N} G^i$.
(e) For any $\underline{\gamma} \in \boldsymbol{\Gamma}$ with $J(\underline{\gamma}) < \infty$ and each $i \in \mathcal{N}$, there exists $u^{i,*} \in \mathsf{U}^i$ such that $J(\underline{\gamma}^{-i}, \gamma^i_{u^{i,*}}) < \infty$, where $\gamma^i_{u^{i,*}} \equiv u^{i,*}$.

Remark 8.8 Note that Assumption 8.2-(d),(e) hold if the cost function is bounded. Therefore, if the static team problem satisfies Assumption 8.1, then Assumption 8.2 (except (b)) holds as well. Hence, the results derived in this section *almost* include the results proved in Section 8.2.1 as a particular case. However, since the analysis in this section is somewhat involved, we presented the compact and bounded case separately.

Our approach can be summarized as follows:

(i) First, we show that there exists a near optimal policy whose range is convex and compact (and thus bounded).
(ii) We prove that the near optimal policy in step (i) can be taken to be continuous.
(iii) We construct a near optimal policy which only uses quantized observations.

(iv) Under one additional assumption, we prove that optimal policies for the finite models constructed using components of the original model can approximate an optimal policy for the original model with arbitrary precision.

The following result states that there exists a near optimal strategy whose range is convex and compact. In what follows, for any subset G of U, we let

$$\boldsymbol{\Gamma}_G := \{\underline{\gamma} \in \boldsymbol{\Gamma} : \underline{\gamma}(\mathsf{Y}) \subset G\}.$$

Lemma 8.9 *Suppose Assumption 8.2-(a),(b),(c),(e) hold. Then, for any $\varepsilon > 0$, there exists a compact subset G of U of the form $G = \prod_{i=1}^N G^i$, where each G^i is convex and compact, such that*

$$\inf_{\underline{\gamma} \in \boldsymbol{\Gamma}_G} J(\underline{\gamma}) < J^* + \varepsilon.$$

Proof Fix any $\varepsilon > 0$. Let $\underline{\gamma} \in \boldsymbol{\Gamma}$ with $J(\underline{\gamma}) < J^* + \varepsilon/2$. We construct the desired G iteratively.

By Assumption 8.2-(e) there exists $u^{1,*} \in \mathsf{U}^1$ such that $J(\underline{\gamma}^{-1}, \gamma^1_{u^{1,*}}) < \infty$. Let $G^1 \subset \mathsf{U}^1$ be a compact set containing $u^{1,*}$. We define

$$\tilde{\gamma}(y^1) = \begin{cases} \gamma^1(y^1), & \text{if } \gamma^1(y^1) \in G^1 \\ u^{1,*}, & \text{otherwise.} \end{cases}$$

Define also $\underline{\gamma}_1 := (\tilde{\gamma}^1, \gamma^2, \ldots, \gamma^N)$, $M_1 := \{y^1 \in \mathsf{Y}^1 : \gamma^1(y^1) \in G^1\}$, and $\tilde{U}^1 = \tilde{\gamma}(Y^1)$. Then, we have

$$\begin{aligned} |J(\underline{\gamma}) - J(\underline{\gamma}_1)| &= \Big| \mathbb{E}\big[c(X, \boldsymbol{Y}, \boldsymbol{U}) 1_{\{Y^1 \in M_1\}}\big] + \mathbb{E}\big[c(X, \boldsymbol{Y}, \boldsymbol{U}) 1_{\{Y^1 \notin M_1\}}\big] \\ &\quad - \mathbb{E}\big[c(X, \boldsymbol{Y}, \boldsymbol{U}^{-1}, \tilde{U}^1) 1_{\{Y^1 \in M_1\}}\big] \\ &\quad - \mathbb{E}\big[c(X, \boldsymbol{Y}, \boldsymbol{U}^{-1}, \tilde{U}^1) 1_{\{Y^1 \notin M_1\}}\big] \Big| \\ &\le \mathbb{E}\big[c(X, \boldsymbol{Y}, \boldsymbol{U}) 1_{\{Y^1 \notin M_1\}}\big] + \mathbb{E}\big[c(X, \boldsymbol{Y}, \boldsymbol{U}^{-1}, \tilde{U}^1) 1_{\{Y^1 \notin M_1\}}\big] \\ &= \int_{\mathsf{X} \times \mathsf{Y} \times \mathsf{U}^{-1} \times (G^1)^c} c(x, \boldsymbol{y}, \boldsymbol{u}) \, \delta_{\underline{\gamma}}(d\boldsymbol{u}) \, \mathbb{P}(dx, d\boldsymbol{y}) \\ &\quad + \int_{\mathsf{X} \times \mathsf{Y} \times \mathsf{U}^{-1} \times (G^1)^c} c(x, \boldsymbol{y}, \boldsymbol{u}) \, \delta_{(\underline{\gamma}^{-1}, \gamma^1_{u^{1,*}})}(d\boldsymbol{u}) \, \mathbb{P}(dx, d\boldsymbol{y}), \end{aligned}$$

where D^c denotes the complement of the set D and $\mathsf{U}^{-1} = \prod_{i=2}^N \mathsf{U}^i$. Recall that $J(\underline{\gamma}^{-1}, \gamma^1_{u^{1,*}}) < \infty$ by Assumption 8.2-(e). Hence, the last expression can be made smaller than $\frac{\varepsilon}{2N}$ by properly choosing G^1 since U^1 is a Borel space [100, Theorem 3.2]. Since the closed convex hull of the set G^1 is compact by Theorem 8.3, we can indeed assume that G^1 is convex without loss of generality. By replacing $\underline{\gamma}$ with $\underline{\gamma}_1$ and applying the same method as above, we can obtain $\underline{\gamma}_2$, and a convex and compact $G^2 \subset \mathsf{U}^2$ such that $|J(\underline{\gamma}_1) - J(\underline{\gamma}_2)| \leq \frac{\varepsilon}{2N}$ and $\gamma^2(\mathsf{Y}^2) \subset G^2$.

Continuing this way, we obtain $G = \prod_{i=1}^N G^i$ and $\underline{\gamma}_N \in \boldsymbol{\Gamma}_G$ such that $\big|J(\underline{\gamma}) - J(\underline{\gamma}_N)\big| < \varepsilon/2$, where G^i is convex and compact for all $i = 1, \ldots, N$. Hence, we have $J(\underline{\gamma}_N) < J^* + \varepsilon$, completing the proof. □

Recall that $\boldsymbol{\Gamma}_c$ denotes the set of continuous strategies. For any $G \subset \mathsf{U}$, we define $\boldsymbol{\Gamma}_{c,G} := \boldsymbol{\Gamma}_c \cap \boldsymbol{\Gamma}_G$; that is, $\boldsymbol{\Gamma}_{c,G}$ is the set of continuous strategies having range in G.

Proposition 8.10 *Suppose Assumption 8.2 holds. Then, for any $\varepsilon > 0$, there exists a compact subset G of U of the form $G = \prod_{i=1}^N G^i$, where each G^i is convex and compact, such that*

$$\inf_{\underline{\gamma} \in \boldsymbol{\Gamma}_{c,G}} J(\underline{\gamma}) < J^* + \varepsilon.$$

Proof Fix any $\varepsilon > 0$. By Lemma 8.9, there exists a compact subset $G = \prod_{i=1}^N G^i$ of U, where G^i is convex and compact, and $\underline{\gamma} \in \boldsymbol{\Gamma}_G$ such that

$$J(\underline{\gamma}) < J^* + \frac{\varepsilon}{2}.$$

Recall that μ^i denotes the distribution of Y^i.

Let δ be positive constant which will be specified later. Analogous to the proof of Proposition 8.4, we construct a continuous strategy $\underline{\tilde{\gamma}}$ which is *almost* equal to $\underline{\gamma}$. For each i, Lusin's theorem implies the existence of a closed set $F_{i,\delta} \subset \mathsf{Y}^i$ such that $\mu^i\big(\mathsf{Y}^i \setminus F_{i,\delta}\big) < \delta$ and the restriction, denoted by π^i, of γ^i on $F_{i,\delta}$ is continuous. By the Dugundji extension theorem there exists a continuous extension $\tilde{\gamma}^i : \mathsf{Y}^i \to G^i$ of π^i. Therefore, $\underline{\tilde{\gamma}} = (\tilde{\gamma}^1, \ldots, \tilde{\gamma}^N) \in \boldsymbol{\Gamma}_{c,G}$. Let $F_\delta := \prod_{i=1}^N F_{i,\delta}$. Then, we have

$$\begin{aligned}\mathbb{P}\big(\mathsf{X} \times (\mathsf{Y} \setminus F_\delta)\big) &\leq \sum_{i=1}^N \mathbb{P}\big(\mathsf{X} \times \mathsf{Y}^{-i} \times (\mathsf{Y}^i \setminus F_{i,\delta})\big) \\ &\leq \sum_{i=1}^N \delta = N\delta.\end{aligned}$$

Recalling the definition of w_G in Assumption 8.2-(d), we have

$$
\begin{aligned}
\left|J(\underline{\gamma}) - J(\underline{\tilde{\gamma}})\right| &= \left| \int_{\mathsf{X}\times\mathsf{Y}} \left[c(x, \mathbf{y}, \underline{\gamma}) - c(x, \mathbf{y}, \underline{\tilde{\gamma}})\right]\mathbb{P}(dx, d\mathbf{y}) \right| \\
&\leq \int_{\mathsf{X}\times(\mathsf{Y}\setminus F)} \left[c(x, \mathbf{y}, \underline{\gamma}) + c(x, \mathbf{y}, \underline{\tilde{\gamma}})\right] \mathbb{P}(dx, d\mathbf{y}) \\
&\leq 2 \int_{\mathsf{X}\times(\mathsf{Y}\setminus F)} w_G(x, \mathbf{y})\, \mathbb{P}(dx, d\mathbf{y}).
\end{aligned}
$$

By Assumption 8.2-(d) w_G is $\mathbb{P}$-integrable so that the measure $w_G(x, \mathbf{y})\mathbb{P}(dx, d\mathbf{y})$ is absolutely continuous with respect to $\mathbb{P}$. Since $\mathbb{P}\big(\mathsf{X} \times (\mathsf{Y} \setminus F_\delta)\big) \to 0$ as $\delta \to 0$, we obtain

$$
\int_{\mathsf{X}\times(\mathsf{Y}\setminus F_\delta)} w_G(x, \mathbf{y})\, \mathbb{P}(dx, d\mathbf{y}) \to 0 \text{ as } \delta \to 0.
$$

Since $J(\underline{\gamma}) < J^* + \frac{\varepsilon}{2}$, there exists a sufficiently small $\delta > 0$ such that $J(\underline{\tilde{\gamma}}) < J^* + \varepsilon$. This completes the proof. □

Since each Y^i is a locally compact separable metric space, there exists a nested sequence of compact sets $\{\mathsf{K}_l^i\}$ such that $\mathsf{K}_l^i \subset \operatorname{int} \mathsf{K}_{l+1}^i$ and $\mathsf{Y}^i = \bigcup_{l=1}^\infty \mathsf{K}_l^i$ [2, Lemma 2.76], where $\operatorname{int} D$ denotes the interior of the set D.

Recall that d_i denotes the metric on Y^i. For each $l \geq 1$, let $\mathsf{Y}_{l,n}^i := \{y_{i,1}, \ldots, y_{i,i_{l,n}}\} \subset \mathsf{K}_l^i$ be an $1/n$-net in K_l^i. Recall that if $\mathsf{Y}_{l,n}^i$ is an $1/n$-net in K_l^i, then for any $y \in \mathsf{K}_l^i$ we have

$$
\min_{z \in \mathsf{Y}_{l,n}^i} d_i(y, z) < \frac{1}{n}.
$$

For each l and n, let $q_{l,n}^i : \mathsf{K}_l^i \to \mathsf{Y}_{l,n}^i$ be the nearest neighbor quantizer; that is, for any $y \in \mathsf{K}_l^i$

$$
q_{l,n}^i(y) = \arg\min_{z \in \mathsf{Y}_{l,n}^i} d_i(y, z),
$$

where ties are broken so that $q_{l,n}^i$ is measurable. If $\mathsf{K}_l^i = [-M, M] \subset \mathsf{Y}^i = \mathbb{R}$ for some $M \in \mathbb{R}_+$, the finite set $\mathsf{Y}_{l,n}^i$ can be chosen such that $q_{l,n}^i$ becomes an uniform quantizer. We let $Q_{l,n}^i : \mathsf{Y}^i \to \mathsf{Y}_{l,n}^i$ denote the extension of $q_{l,n}^i$ to Y^i given by

$$
Q_{l,n}^i(y) := \begin{cases} q_{l,n}^i(y), & \text{if } y \in \mathsf{K}_l^i, \\ y_{i,0}, & \text{otherwise,} \end{cases}
$$

where $y_{i,0} \in \mathsf{Y}^i$ is some auxiliary element. Define $\Gamma_{l,n}^i = \Gamma^i \circ Q_{l,n}^i \subset \Gamma^i$; that is, $\Gamma_{l,n}^i$ is defined to be the set of all strategies $\tilde{\gamma}^i \in \Gamma^i$ of the form $\tilde{\gamma}^i = \gamma^i \circ Q_{l,n}^i$, where $\gamma^i \in \Gamma^i$. Define also $\boldsymbol{\Gamma}_{l,n} := \prod_{i=1}^N \Gamma_{l,n}^i \subset \boldsymbol{\Gamma}$. The following theorem states that an optimal (or almost optimal) policy can be approximated with arbitrarily small approximation error for the induced costs by policies in $\boldsymbol{\Gamma}_{l,n}$ for sufficiently large l and n.

Theorem 8.11 *For any $\varepsilon > 0$, there exist $(l, n(l))$ and $\underline{\gamma} \in \boldsymbol{\Gamma}_{l,n(l)}$ such that*

$$J(\underline{\gamma}) < J^* + \varepsilon.$$

Proof By Proposition 8.10, there exists $\underline{\gamma} \in \boldsymbol{\Gamma}_{c,G}$ such that $J(\underline{\gamma}) < J^* + \varepsilon/4$, where $G = \prod_{i=1}^N G^i$ and each G^i is convex and compact. For each l and n, we define $\gamma_{l,n}^i := \gamma^i \circ Q_{l,n}^i$ and $\underline{\gamma}_{l,n} = (\gamma_{l,n}^1, \ldots, \gamma_{l,n}^N)$. Define also $u^{i,*} := \gamma^i(y_{i,0}) \in G^i$.

Let $\mathcal{N}^*$ denote the collection of all subsets of $\mathcal{N}$ except the empty set. For any $s \in \mathcal{N}^*$, we define

$$u^{s,*} := \left(u^{i,*}\right)_{i \in s}, \ \underline{\gamma}_{u^{s,*}} := \left(\gamma_{u^{i,*}}^i\right)_{i \in s}, \ \underline{\gamma}^{-s} := \left(\gamma^i\right)_{i \notin s},$$

and

$$\mathsf{K}_l^s := \prod_{i \in s} (\mathsf{K}_l^i)^c \times \prod_{i \notin s} \mathsf{K}_l^i.$$

Recall that $\gamma_{u^{i,*}}^i$ is the strategy which maps any $y^i \in \mathsf{Y}^i$ to $u^{i,*}$. Let $\mathsf{K}_l = \prod_{i=1}^N \mathsf{K}_l^i$ and observe that

$$(\mathsf{X} \times \mathsf{K}_l)^c = \mathsf{X} \times \bigcup_{s \in \mathcal{N}^*} \mathsf{K}_l^s.$$

Note that since the range of the strategy $(\underline{\gamma}^{-s}, \underline{\gamma}_{u^{s,*}})$ is contained in G, we have $J(\underline{\gamma}^{-s}, \underline{\gamma}_{u^{s,*}}) \leq \int_{\mathsf{X} \times \mathsf{Y}} w_G(x, \boldsymbol{y}) \mathbb{P}(dx, d\boldsymbol{y}) < \infty$ for all $s \in \mathcal{N}^*$ by Assumption 8.2-(d). Hence, there exists l sufficiently large such that

$$\left| J(\underline{\gamma}) - \int_{\mathsf{X} \times \mathsf{K}_l} c(x, \boldsymbol{y}, \underline{\gamma}) \mathbb{P}(dx, d\boldsymbol{y}) \right| \leq \frac{\varepsilon}{4},$$

and

$$\int_{\mathsf{X} \times \mathsf{K}_l^s} c(x, \boldsymbol{y}, \underline{\gamma}^{-s}, \underline{\gamma}_{u^{s,*}}) \mathbb{P}(dx, d\boldsymbol{y}) \leq \frac{\varepsilon}{2^{N+1}}, \text{ for all } s \in \mathcal{N}^*.$$

Let $q_{l,n}^{-s} = (q_{l,n}^i)_{i \notin s}$. Then, we have

$$\begin{aligned}
\limsup_{n\to\infty} |J(\underline{\gamma}) - J(\underline{\gamma}_{l,n})| &\leq \left| J(\underline{\gamma}) - \int_{\mathsf{X}\times\mathsf{K}_l} c(x, \mathbf{y}, \underline{\gamma})\, d\mathbb{P} \right| \\
&+ \limsup_{n\to\infty} \left| \int_{\mathsf{X}\times\mathsf{K}_l} c(x, \mathbf{y}, \underline{\gamma})\, d\mathbb{P} - \int_{\mathsf{X}\times\mathsf{K}_l} c(x, \mathbf{y}, \underline{\gamma}_{l,n})\, d\mathbb{P} \right| \\
&+ \sum_{s\in\mathcal{N}^*} \limsup_{n\to\infty} \int_{\mathsf{X}\times\mathsf{K}_l^s} c(x, \mathbf{y}, \underline{\gamma}^{-s} \circ q_{l,n}^{-s}, \underline{\gamma}_{u^{s,*}})\, d\mathbb{P} \\
&\leq \frac{\varepsilon}{4} + \limsup_{n\to\infty} \left| \int_{\mathsf{X}\times\mathsf{K}_l} c(x, \mathbf{y}, \underline{\gamma}) d\mathbb{P} - \int_{\mathsf{X}\times\mathsf{K}_l} c(x, \mathbf{y}, \underline{\gamma}_{l,n}) d\mathbb{P} \right| \\
&+ \sum_{s\in\mathcal{N}^*} \limsup_{n\to\infty} \left| \int_{\mathsf{X}\times\mathsf{K}_l^s} c(x, \mathbf{y}, \underline{\gamma}^{-s} \circ q_{l,n}^{-s}, \underline{\gamma}_{u^{s,*}})\, d\mathbb{P} \right. \\
&\left. - \int_{\mathsf{X}\times\mathsf{K}_l^s} c(x, \mathbf{y}, \underline{\gamma}^{-s}, \underline{\gamma}_{u^{s,*}})\, d\mathbb{P} \right| \\
&+ \sum_{s\in\mathcal{N}^*} \int_{\mathsf{X}\times\mathsf{K}_l^s} c(x, \mathbf{y}, \underline{\gamma}^{-s}, \underline{\gamma}_{u^{s,*}})\, d\mathbb{P} \\
&\leq \frac{\varepsilon}{4} + \sum_{s\in\mathcal{N}^*} \frac{\varepsilon}{2^{N+1}} + \limsup_{n\to\infty} \left| \int_{\mathsf{X}\times\mathsf{K}_l} c(x, \mathbf{y}, \underline{\gamma})\, d\mathbb{P} - \int_{\mathsf{X}\times\mathsf{K}_l} c(x, \mathbf{y}, \underline{\gamma}_{l,n})\, d\mathbb{P} \right| \\
&+ \sum_{s\in\mathcal{N}^*} \limsup_{n\to\infty} \left| \int_{\mathsf{X}\times\mathsf{K}_l^s} c(x, \mathbf{y}, \underline{\gamma}^{-s} \circ q_{l,n}^{-s}, \underline{\gamma}_{u^{s,*}})\, d\mathbb{P} \right. \\
&\left. - \int_{\mathsf{X}\times\mathsf{K}_l^s} c(x, \mathbf{y}, \underline{\gamma}^{-s}, \underline{\gamma}_{u^{s,*}})\, d\mathbb{P} \right|.
\end{aligned}$$

Note that in the last expression, the integrands in the third and fourth terms are upper bounded by w_G. Since w_G is $\mathbb{P}$-integrable by Assumption 8.2-(d), $\gamma^i \circ q_{l,n}^i \to \gamma^i$ on K_l^i as $n \to \infty$ (recall that γ^i is continuous), and c is continuous in $\boldsymbol{u}$, the third and fourth terms in the last expression converge to zero as $n \to \infty$ by the dominated convergence theorem. Hence, there exists a sufficiently large $n(l)$ such that the last expression becomes less than $3\varepsilon/4$. Therefore, $J(\underline{\gamma}_{l,n(l)}) < J^* + \varepsilon$, completing the proof. □

The above result implies that to compute a near optimal policy for the team problem it is sufficient to choose a strategy based on the quantized observations $(Q_{l,n}^1(Y^1), \ldots, Q_{l,n}^N(Y^N))$ for sufficiently large l and n. Furthermore, this nearly optimal strategy can have a compact range of the form $G = \prod_{i=1}^N G^i$, where G^i is convex and compact for each $i = 1, \ldots, N$. However, to obtain a result analogous to the Theorem 8.7, we need to impose a further assumption. To this end, we first introduce a finite observation model. For each (l, n), let $\mathsf{Z}^i_{l,n} := \{y_{i,0}, y_{i,1}, \dots, y_{i,i_{l,n}}\}$ (i.e., the output levels of $Q^i_{l,n}$) and define the stochastic kernels $W^i_{l,n}(\,\cdot\,|x)$ on $\mathsf{Z}^i_{l,n}$ given X as follows:

$$W^i_{l,n}(\,\cdot\,|x) := \sum_{j=0}^{i_{l,n}} W(\mathscr{S}^{l,n}_{i,j}|x)\delta_{y_{i,j}}(\,\cdot\,),$$

where $\mathscr{S}^{l,n}_{i,j} := \{y \in \mathsf{Y}^i : Q^i_{l,n}(y) = y_{i,j}\}$. Let $\Pi^i_{n,l} := \{\pi^i : \mathsf{Z}^i_{l,n} \to \mathsf{U}^i, \pi^i \text{ measurable}\}$ and $\boldsymbol{\Pi}_{l,n} := \prod_{i=1}^N \Pi^i_{l,n}$. Define $J_{l,n} : \boldsymbol{\Pi}_{l,n} \to [0, \infty)$ as

$$J_{l,n}(\underline{\pi}) := \int_{\mathsf{X}\times\mathsf{Z}_{l,n}} c(x, \boldsymbol{y}, \boldsymbol{u})\mathbb{P}_{l,n}(dx, d\boldsymbol{y}),$$

where $\underline{\pi} = (\pi^1, \dots, \pi^N)$, $\boldsymbol{u} = \underline{\pi}(\boldsymbol{y})$, $\mathsf{Z}_{l,n} = \prod_{i=1}^N \mathsf{Z}^i_{l,n}$, and

$$\mathbb{P}_{l,n}(dx, d\boldsymbol{y}) = \mathbb{P}(dx)\prod_{i=1}^N W^i_{l,n}(dy^i|x).$$

Note that the probability measure $\mathbb{P}_{l,n}$ can also be treated as a measure on $\mathsf{X} \times \mathsf{Y}$. In this case, it is not difficult to prove that $\mathbb{P}_{l,n}$ converges to $\mathbb{P}$ weakly as $l, n \to \infty$. For any compact subset G of U, we also define $\boldsymbol{\Pi}^G_{l,n} := \{\underline{\pi} \in \boldsymbol{\Pi}_{l,n} : \underline{\pi}(\mathsf{Z}_{l,n}) \subset G\}$.

Assumption 8.3 *For any compact subset G of U of the form $G = \prod_{i=1}^N G^i$, we assume that the function w_G is uniformly integrable with respect to the measures $\{\mathbb{P}_{l,n}\}$; that is,*

$$\lim_{R\to\infty} \sup_{l,n} \int_{\{w_G>R\}} w_G(x, \boldsymbol{y})\, d\mathbb{P}_{l,n} = 0.$$

Lemma 8.12 *Let $\{\underline{\pi}_{l,n}\}$ be a sequence of strategies such that $\underline{\pi}_{l,n} \in \boldsymbol{\Pi}^G_{l,n}$, where $G = \prod_{i=1}^N G^i$ and each G^i is convex and compact. For each l and n, define $\underline{\gamma}_{l,n} := \underline{\pi}_{l,n} \circ Q_{l,n}$, where $Q_{l,n} := (Q^1_{l,n}, \dots, Q^N_{l,n})$. Then, we have*

$$\lim_{l,n\to\infty} |J_{l,n}(\underline{\pi}_{l,n}) - J(\underline{\gamma}_{l,n})| = 0.$$

Proof Let us introduce the following finite measures on $\mathsf{X} \times \mathsf{Y}$:

$$\mu_G(S) := \int_S w_G(x, \boldsymbol{y})\, d\mathbb{P},$$

$$\mu_G^{l,n}(S) := \int_S w_G(x, \mathbf{y})\, d\mathbb{P}_{l,n}.$$

Since $\mathbb{P}_{l,n}$ converges to $\mathbb{P}$ weakly, by [122, Theorem 3.5] and Assumption 8.3 we have $\mu_G^{l,n} \to \mu_G$ weakly as $l, n \to \infty$. Hence, the sequence $\{\mu_G^{l,n}\}$ is tight. Therefore, there exists a compact subset K of $\mathsf{X} \times \mathsf{Y}$ such that $\mu_G(K^c) < \varepsilon/2$ and $\mu_G^{l,n}(K^c) < \varepsilon/2$ for all l, n. Then, we have

$$\begin{aligned}
|J_{l,n}(\underline{\pi}_{l,n}) - J(\underline{\gamma}_{l,n})| &= \left| \int_{\mathsf{X}\times\mathsf{Y}} c(x, Q_{l,n}(\mathbf{y}), \underline{\gamma}_{l,n})\, d\mathbb{P} - \int_{\mathsf{X}\times\mathsf{Y}} c(x, \mathbf{y}, \underline{\gamma}_{l,n})\, d\mathbb{P} \right| \\
&\leq \int_K \left| c(x, Q_{l,n}(\mathbf{y}), \underline{\gamma}_{l,n}) - c(x, \mathbf{y}, \underline{\gamma}_{l,n}) \right| d\mathbb{P} \\
&\quad + \int_{K^c} w_G(x, \mathbf{y})\, d\mathbb{P} + \int_{K^c} w_G(x, \mathbf{y})\, d\mathbb{P}_{l,n}.
\end{aligned}$$

The first term in the last expression goes to zero as $l, n \to \infty$ by the dominated convergence theorem and the fact that c is bounded and continuous in $\mathbf{y}$. The second term is less than ε. Since ε is arbitrary, this completes the proof. □

The following theorem, which is a consequence of Theorem 8.11, is the main result of this section. It states that to compute a near optimal strategy for the original team problem, it is sufficient to compute an optimal (or an almost optimal) policy for the team problem described above.

Theorem 8.13 *For any $\varepsilon > 0$, there exists a pair $(l, n(l))$ and a compact subset $G = \prod_{i=1}^N G^i$ of U such that the optimal (or almost optimal) policy $\underline{\pi}^*$ in the set $\boldsymbol{\Pi}_{l,n(l)}^G$ for the cost $J_{l,n(l)}$ is ε-optimal for the original team problem when $\underline{\pi}^* = (\pi^{1*}, \ldots, \pi^{N*})$ is extended to Y via $\gamma^i = \pi^{i*} \circ Q_{l,n(l)}^i$.*

Proof Fix any $\varepsilon > 0$. By Lemma 8.9 and Theorem 8.11, there exists compact subset G of U of the form $G = \prod_{i=1}^N G^i$ such that

$$\lim_{l,n\to\infty} \inf_{\underline{\gamma}\in\boldsymbol{\Gamma}_{l,n}\cap\boldsymbol{\Gamma}_G} J(\underline{\gamma}) - J^* < \varepsilon.$$

For each $l, n \geq 1$, let $\underline{\gamma}_{l,n} \in \boldsymbol{\Gamma}_{l,n} \cap \boldsymbol{\Gamma}_G$ be such that $J(\underline{\gamma}_{l,n}) < \inf_{\underline{\gamma}\in\boldsymbol{\Gamma}_{l,n}\cap\boldsymbol{\Gamma}_G} J(\underline{\gamma}) + 1/(n+l)$. Define $\underline{\pi}_{l,n}$ as the restriction of $\underline{\gamma}_{l,n}$ to the set $\mathsf{Z}_{l,n}$. Then, we have

$$\begin{aligned}
J^* + \varepsilon &\geq \lim_{l,n\to\infty} J(\underline{\gamma}_{l,n}) \\
&= \lim_{l,n\to\infty} J_{l,n}(\underline{\pi}_{l,n}) \text{ (by Lemma 8.12)} \\
&\geq \limsup_{l,n\to\infty} \inf_{\underline{\pi}\in\boldsymbol{\Pi}_{l,n}^G} J_{l,n}(\underline{\pi}).
\end{aligned}$$

For the reverse inequality, for each $l, n \geq 1$, let $\underline{\pi}_{l,n} \in \boldsymbol{\Pi}_{l,n}^G$ be such that $J_{l,n}(\underline{\pi}_{l,n}) < \inf_{\underline{\pi} \in \boldsymbol{\Pi}_{l,n}^G} J_{l,n}(\underline{\pi}) + 1/(n+l)$. Define $\underline{\gamma}_{l,n} := \underline{\pi}_{l,n} \circ Q_{l,n}$. Then, we have

$$\begin{aligned}
\liminf_{l,n \to \infty} \inf_{\underline{\pi} \in \boldsymbol{\Pi}_{l,n}^G} J_{l,n}(\underline{\pi}) &= \liminf_{l,n \to \infty} J_{l,n}(\underline{\pi}_{l,n}) \\
&= \liminf_{l,n \to \infty} J(\underline{\gamma}_{l,n}) \text{ (by Lemma 8.12)} \\
&\geq J^*.
\end{aligned}$$

This completes the proof. □

8.3 Approximation of Dynamic Team Problems

The results for the static case apply also to the dynamic case, through a static reduction, as reviewed in Section 7.2.

8.3.1 Approximation of Dynamic Teams Admitting a Static Reduction

The next theorem is the main result of this section. It states that for a class of dynamic team problems, finite models can approximate an optimal policy with arbitrary precision.

Theorem 8.14 *Suppose Assumptions 8.2-(a),(b),(c),(e) and 7.1 hold. In addition, suppose that for each $i = 1, \ldots, N$, $f_i(w, \boldsymbol{u}^{i-1}, y^i)$ is continuous in $\boldsymbol{u}^{i-1}$ and y^i, and*

$$\sup_{\boldsymbol{u} \in G} c(w, \boldsymbol{y}, \boldsymbol{u}) \prod_{i=1}^{N} f_i(w, \boldsymbol{u}^{i-1}, y^i) \text{ is } \tilde{\mathbb{P}}_{l,n} - \text{uniformly integrable}$$

for all compact $G \subset \mathsf{U}$ of the form $G = \prod_{i=1}^N G^i$. Then, the static reduction of the dynamic team model satisfies Assumptions 8.2 and 8.3. Therefore, Theorems 8.11 and 8.13 hold for the dynamic team problem. In particular, Theorems 8.11 and 8.13 hold for the dynamic team problems satisfying Assumptions 8.2, 8.3, and 7.1, if f_i is bounded and continuous in $\boldsymbol{u}^{i-1}$ and y^i for each $i = 1, \ldots, N$.

It is worth noting that for partially nested (or quasi-classical) information structures with a static reduction, the cost function does not change as a result of the static reduction, unlike in the static reduction for general nonclassical systems in Section 7.2; see [150] for further discussions. Therefore, if the partially nested dynamic team satisfies Assumption 8.2, then its static reduction also satisfies it. Hence, Theorems 8.11 and 8.13 hold for such problems.

8.4 Discretization of the Action Spaces

For computing near optimal strategies for static team problems using numerical algorithms, the action spaces U^i must be finite. In this section, we show that the action spaces can be taken to be finite in the finite observation models if a sufficiently large number of points are used for accurate approximation. In this section, we consider the most general case studied in Section 8.2.2.

We note that the results derived in this section can be applied to dynamic teams which admit static reduction and satisfy conditions in Theorem 8.14.

Recall the finite observation models constructed in Section 8.2.2. For each (l, n), the finite model have the following components: $\{\mathsf{X}, \mathsf{Z}_{l,n}^i, \mathsf{U}^i, W_{l,n}^i(\,\cdot\,|x), c, \mathbb{P}, i \in \mathcal{N}\}$, where X is the state space, $\mathsf{Z}_{l,n}^i$ is the observation space for Agent i, U^i is the action space for Agent i, $W_{l,n}^i(\,\cdot\,|x)$ observation channel from state to the observation of Agent i, c is the cost function, and $\mathbb{P}$ is the distribution of the state. Furthermore, the strategy spaces are defined as $\Pi_{n,l}^i := \{\pi^i : \mathsf{Z}_{l,n}^i \to \mathsf{U}^i, \pi^i \text{ measurable}\}$ and $\boldsymbol{\Pi}_{l,n} := \prod_{i=1}^N \Pi_{l,n}^i$. Then the cost function $J_{l,n} : \boldsymbol{\Pi}_{l,n} \to [0, \infty)$ is given by

$$J_{l,n}(\underline{\pi}) := \int_{\mathsf{X}\times\mathsf{Z}_{l,n}} c(x, \mathbf{y}, \boldsymbol{u})\mathbb{P}_{l,n}(dx, d\mathbf{y}),$$

where $\underline{\pi} = (\pi^1, \ldots, \pi^N)$, $\boldsymbol{u} = \underline{\pi}(\mathbf{y})$, $\mathsf{Z}_{l,n} = \prod_{i=1}^N \mathsf{Z}_{l,n}^i$, and $\mathbb{P}_{l,n}(dx, d\mathbf{y}) = \mathbb{P}(dx)\prod_{i=1}^N W_{l,n}^i(dy^i|x)$.

The theorem below is the main result of this section which states that one can approximate optimal strategy in $\boldsymbol{\Pi}_{l,n}$ by strategies taking values in a finite set.

Theorem 8.15 *Suppose that the original static team problem satisfies Assumptions 8.2 and 8.3. Then, for each (l, n) and for any $\varepsilon > 0$, there exist finite sets $\mathsf{U}_\varepsilon^i \subset \mathsf{U}^i$ for $i = 1, \ldots, N$ such that*

$$\inf_{\underline{\pi}\in\boldsymbol{\Pi}_{l,n}^\varepsilon} J_{l,n}(\underline{\pi}) < J_{l,n}^* + \varepsilon,$$

where $\boldsymbol{\Pi}_{l,n}^\varepsilon := \{\underline{\pi} \in \boldsymbol{\Pi}_{l,n} : \pi^i(\mathsf{Z}_{l,n}^i) \subset \mathsf{U}_\varepsilon^i, i \in \mathcal{N}\}$.

Proof Fix any (l, n) and ε. Let us choose $\underline{\pi}^\varepsilon \in \boldsymbol{\Pi}_{l,n}$ such that

$$J_{l,n}(\underline{\pi}^\varepsilon) < \inf_{\underline{\pi}\in\boldsymbol{\Pi}_{l,n}} J_{l,n}(\underline{\pi}) + \frac{\varepsilon}{2}.$$

Note that for any $i = 1, \ldots, N$, the range of $\pi^{i,\varepsilon}$ is a finite subset of U^i and so, is contained in some compact and convex subset G^i of U^i. Define $G = \prod_{i=1}^N G^i$.

Let ρ_i denote the metric on U^i. Since G^i is compact, one can find a finite set $\mathsf{U}_k^i := \{u_{i,1}, \ldots, u_{i,i_k}\} \subset G^i$ which is a $1/k$-net in G^i. Define $\Pi_k^i := \{\pi^i \in \Pi_{l,n}^{G^i} : \pi^i(\mathsf{Z}_{l,n}^i) \subset \mathsf{U}_k^i\}$ and $\boldsymbol{\Pi}_k = \prod_{i=1}^N \Pi_k^i$. For $\pi^{i,\varepsilon}$, we let

$$\pi_k^{i,\varepsilon}(y) := \arg\min_{u \in \mathsf{U}_k^i} \rho_i(\pi^{i,\varepsilon}(y), u).$$

Hence

$$\sup_{y \in \mathsf{Z}_{l,n}^i} \rho_i(\pi^{i,\varepsilon}(y), \pi_k^{i,\varepsilon}(y)) < 1/k. \tag{8.1}$$

We define $\underline{\pi}_k^{\varepsilon} = (\pi_k^{1,\varepsilon}, \ldots, \pi_k^{N,\varepsilon})$. Then we have

$$\begin{aligned}
\inf_{\underline{\pi} \in \boldsymbol{\Pi}_k} J_{l,n}(\underline{\pi}) - \inf_{\underline{\pi} \in \boldsymbol{\Pi}_{l,n}} J_{l,n}(\underline{\pi}) &< \inf_{\underline{\pi} \in \boldsymbol{\Pi}_k} J_{l,n}(\underline{\pi}) - J_{l,n}(\underline{\pi}^{\varepsilon}) + \frac{\varepsilon}{2} \\
&\leq J_{l,n}(\underline{\pi}_k^{\varepsilon}) - J_{l,n}(\underline{\pi}^{\varepsilon}) + \frac{\varepsilon}{2} \\
&\leq \int_{\mathsf{X} \times \mathsf{Z}_{l,n}} \left| c(x, \mathbf{y}, \underline{\pi}_k^{\varepsilon}) - c(x, \mathbf{y}, \underline{\pi}^{\varepsilon}) \right| d\mathbb{P}_{l,n} + \frac{\varepsilon}{2}.
\end{aligned}$$

The last integral converges to zero as $k \to \infty$ by the dominated convergence theorem since: (i) $c(x, \mathbf{y}, \underline{\pi}_k^{\varepsilon}) \to c(x, \mathbf{y}, \underline{\pi}^{\varepsilon})$ as $k \to \infty$ by (8.1) and Assumption 8.2-(a), (ii) $c(x, \mathbf{y}, \underline{\pi}_k^{\varepsilon}), c(x, \mathbf{y}, \underline{\pi}^{\varepsilon}) \leq w_G(x, \mathbf{y})$ for all $(x, \mathbf{y}) \in \mathsf{X} \times \mathsf{Z}_{l,n}$, and (iii) w_G is $\mathbb{P}_{l,n}$-integrable. Therefore, there exists k_0 sufficiently large such that the last expression is less than ε. By choosing $\mathsf{U}_{\varepsilon}^i = \mathsf{U}_{k_0}^i$, for $i = 1, \ldots, N$, the proof is complete. □

8.5 On Rates of Convergence in Quantized Approximations

Unlike the results in Part I of the book, it is in general not possible to obtain a priori convergence guarantees for decentralized stochastic control problems.

One exception is when optimal policies are proved to be Lipschitz in their measurement variables or the decision makers are restricted to adopt policies which are Lipschitz in their measurement variables. In this case, if the observations spaces are compact and the cost function is bounded and Lipschitz in observation and action variables, then one can obtain rate of convergence result in terms of the number of grid points, the Lipschitz constant of optimal policy and the Lipschitz constant of the cost function. To obtain a rate of convergence analysis for the unbounded case, in addition to these assumptions, one may need to impose some density assumption on the joint distribution of the state and observations which has also sufficient smoothness.

One method to prove the optimality of policies which are Lipschitz in their measurement variables is given in [54]. In this method, under some convexity assumptions it is first proved that person-by-person optimal policies are also globally optimal. Then, a minimizing sequence is constructed using best response maps of each agent. If the functions that are minimized in these best response maps are smooth enough, then one can deduce the differentiability of the optimal policies via the implicit function theorem. If these differentials are bounded, then it is straightforward to conclude that optimal policies are Lipschitz. However, in order to carry out such an analysis, one needs to impose restrictive assumptions on the observation and action spaces, observation kernels, and the cost function as can be seen in [54].

8.6 Concluding Remarks

In this chapter, approximation of both static and dynamic team problems by finite models was considered. Under mild technical conditions, we showed that the finite model obtained by quantizing uniformly the observation and action spaces on finite grids provides a near optimal strategy if the number of grid points is sufficiently large. The results of this chapter build on [120].

For a further reading, we refer the reader to [52, 54] in which approximations of a class of static team problems are considered under fairly strong assumptions. In these papers, in addition to asymptotic convergence result, error bounds on the accuracy of near optimal solutions are established, where near optimal strategies are expressed as linear combinations of basis functions with adjustable parameters.

Chapter 9
Asymptotic Optimality of Finite Models for Witsenhausen's Counterexample and Beyond

9.1 Introduction

In this chapter, we study the approximation of Witsenhausen's counterexample and the Gaussian relay channel problem by using the results of the previous chapter. In particular, our goal is to establish that finite models obtained through the uniform quantization of the observation and action spaces result in a sequence of policies whose costs converge to the value function. We note that the operation of quantization has typically been the method to show that a nonlinear policy can perform better than an optimal linear policy, for both Witsenhausen's counterexample [10, 86] and the Gaussian relay channel problem [88, 152]. Our findings show that for a large class of problems, quantized policies not only may perform better than linear policies, but that they are actually almost optimal.

9.2 Witsenhausen's Counterexample

9.2.1 Witsenhausen's Counterexample and Its Static Reduction

In Witsenhausen's celebrated counterexample [138], depicted in Figure. 9.1, there are two decision makers: Agent 1 observes a zero mean and unit variance Gaussian random variable Y^1 and decides its strategy U^1. Agent 2 observes $Y^2 := U^1 + V$, where V is a standard (zero mean and unit variance) Gaussian noise independent of Y^1, and decides its strategy U^2.

N. Saldi et al., *Finite Approximations in Discrete-Time Stochastic Control*,
Systems & Control: Foundations & Applications,
https://doi.org/10.1007/978-3-319-79033-6_9

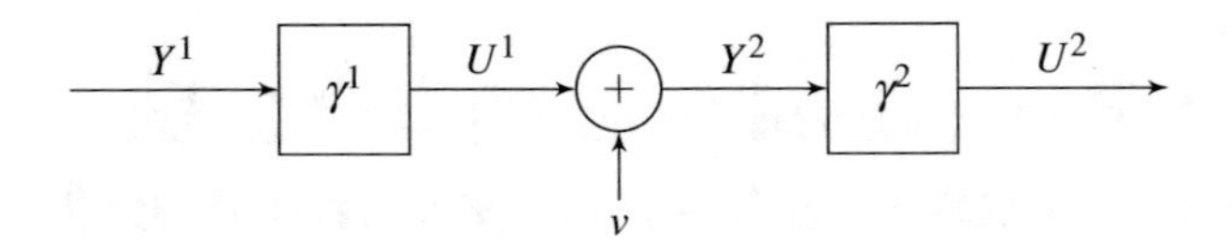

Fig. 9.1 Witsenhausen's counterexample

The cost function of the team is given by

$$c(y^1, u^1, u^2) = l(u^1 - y^1)^2 + (u^2 - u^1)^2,$$

where $l \in \mathbb{R}_+$. In this decentralized system, the state of the nature X can be regarded as a degenerate (constant) random variable. Let $g(y) := \frac{1}{\sqrt{2\pi}} \exp\{-y^2/2\}$. Then we have

$$W^2(S|u^1) = \int_S g(y^2 - u^1) m(dy^2),$$

where m denotes the Lebesgue measure. Let

$$f(u^1, y^2) := \exp\Big\{ - \frac{(u^1)^2 - 2y^2u^1}{2} \Big\} \tag{9.1}$$

so that $g(y^2 - u^1) = f(u^1, y^2) \frac{1}{\sqrt{2\pi}} \exp\{-(y^2)^2/2\}$. The static reduction proceeds as follows: for any policy $\underline{\gamma}$, we have

$$J(\underline{\gamma}) = \int c(y^1, u^1, u^2) W^2(dy^2|u^1) \delta_{\gamma^1(y^1)}(du^1) \mathbb{P}_{\mathrm{g}}(dy^1)$$

$$= \int c(y^1, u^1, u^2) f(u^1, y^2) \mathbb{P}_{\mathrm{g}}(dy^2) \mathbb{P}_{\mathrm{g}}(dy^1),$$

where $\mathbb{P}_{\mathrm{g}}$ denotes the standard Gaussian distribution. Hence, by defining $\tilde{c}(y^1, y^2, u^1, u^2) = c(y^1, u^1, u^2) f(u^1, y^2)$ and $\widetilde{\mathbb{P}}(dy^1, dy^2) = \mathbb{P}_{\mathrm{g}}(dy^1)\mathbb{P}_{\mathrm{g}}(dy^2)$, we can write $J(\underline{\gamma})$ as

$$J(\underline{\gamma}) = \int \tilde{c}(y^1, y^2, u^1, u^2) \widetilde{\mathbb{P}}(dy^1, dy^2). \tag{9.2}$$

Therefore, in the static reduction of Witsenhausen's counterexample, the agents observe independent zero mean and unit variance Gaussian random variables.

Before proceeding to the next section, we state an auxiliary result which will be used in the sequel.

Lemma 9.1 *Let f and g be nonnegative real functions defined on metric spaces E_1 and E_2, respectively. Suppose*

$$\lim_{R\to\infty} \sup_{n\geq 1} \int_{\{f>R\}} f \, d\mu_n = 0$$

$$\lim_{R\to\infty} \sup_{n\geq 1} \int_{\{g>R\}} g \, d\nu_n = 0$$

for some sequence of probability measures $\{\mu_n\}$ and $\{\nu_n\}$. Then, we have

$$\lim_{R\to\infty} \sup_{n} \int_{\{fg>R\}} f(e_1)g(e_2) \, d\mu_n \otimes d\nu_n = 0.$$

9.2.2 Approximation of Witsenhausen's Counterexample

To tackle approximation problem for Witsenhausen's counterexample we show that the conditions in Theorem 8.14 hold and therefore, Theorems 8.11 and 8.13 can be applied.

The cost function of the static reduction is given by

$$\tilde{c}(y^1, y^2, u^1, u^2) = \big(l(u^1 - y^1)^2 + (u^2 - u^1)^2\big) f(u^1, y^2),$$

where $f(u^1, y^2)$ is given in (9.1). Note that the strategy spaces of the original problem and its static reduction are identical, and same strategies induce the same team costs.

Lemma 9.2 *For any $(\gamma^1, \gamma^2) \in \Gamma^1 \times \Gamma^2$ with $J(\gamma^1, \gamma^2) < \infty$, we have $\mathbb{E}\big[\gamma^i(Y^i)^2\big] < \infty$, for $i = 1, 2$.*

Proof To prove the lemma, we use the original problem setup instead of its static reduction. Fix any strategy (γ^1, γ^2) with finite cost. Since $E\big[(U^1 - Y^1)^2\big] < \infty$ and $U^1 = (U^1 - Y^1) + Y^1$, we have $\mathbb{E}\big[(U^1)^2\big] < \infty$. We also have $\mathbb{E}\big[(U^2)^2\big] < \infty$, completing the proof. □

For any $l \in \mathbb{R}_+$, let $L := [-l, l]$ and let $q_{l,n}$ denote the uniform quantizer on L having n output levels; that is,

$$q_{l,n} : L \to \{y_1, \ldots, y_n\} \subset L$$

where $y_j = -l + (j - \frac{1}{2})\Delta$, $j = 1, \ldots, n$ and

$$q^{-1}(y_j) = \Big[y_j - \frac{\Delta}{2}, y_j + \frac{\Delta}{2}\Big),$$

where $\Delta = \frac{2l}{n}$. Let us extend $q_{l,n}$ to $\mathbb{R}$ by mapping $\mathbb{R} \setminus L$ to $y_0 = 0$. For each (l, n), let $\mathsf{Z}_{l,n} := \{y_0, y_1, \ldots, y_n\}$ (i.e., output levels of the extended $q_{l,n}$) and define the probability measure $\mathbb{P}_{l,n}$ on $\mathsf{Z}_{l,n}$ as

$$\mathbb{P}_{l,n}(y_i) = \mathbb{P}_{\mathsf{g}}(q_{l,n}^{-1}(y_i)).$$

Moreover, let $\Pi_{l,n}^i := \{\pi^i : \mathsf{Z}_{l,n} \to \mathsf{U}^i,\ \pi^i \text{ measurable}\}$ and define

$$J_{l,n}(\pi^1, \pi^2) := \sum_{j,i=0}^{n} \tilde{c}(y_i, y_j, \pi^1(y_i), \pi^2(y_j))\mathbb{P}_{l,n}(y_i)\mathbb{P}_{l,n}(y_j).$$

With the help of Lemma 9.2, we now prove the following result.

Proposition 9.3 *Witsenhausen's counterexample satisfies conditions in Theorem 8.14.*

Proof Assumption 8.2-(a),(b),(c) and Assumption 7.1 clearly hold. To prove Assumption 8.2-(e), we introduce the following notation. For any strategy (γ^1, γ^2), we let $\mathbb{E}_{\gamma^1,\gamma^2}$ denote the corresponding expectation operation. Pick (γ^1, γ^2) with $J(\gamma^1, \gamma^2) < \infty$. Since

$$\mathbb{E}_{\gamma^1,\gamma^2}\Big[\mathbb{E}_{\gamma^1,\gamma^2}\Big[(U^1 - U^2)^2 \Big| U^1\Big]\Big] = \mathbb{E}_{\gamma^1,\gamma^2}\Big[(U^1 - U^2)^2\Big]$$

by the law of total expectation, there exists $u^{1,*} \in \mathsf{U}^1$ such that $\mathbb{E}_{\gamma^1,\gamma^2}\big[(U^1 - U^2)^2 \big| U^1 = u^{1,*}\big] < \infty$. Let $u^{2,*} = 0$. Then we have

$$\begin{aligned} J(\gamma^1, \gamma^2_{u^{2,*}}) &= \mathbb{E}_{\gamma^1,\gamma^2_{u^{2,*}}}\Big[l(U^1 - Y^1)^2 + (U^1)^2\Big] \\ &= \mathbb{E}_{\gamma^1,\gamma^2}\Big[l(U^1 - Y^1)^2 + (U^1)^2\Big] < \infty \text{ (by Lemma 9.2)} \end{aligned}$$

and

$$\begin{aligned} J(\gamma^1_{u^{1,*}}, \gamma^2) &= \mathbb{E}_{\gamma^1_{u^{1,*}},\gamma^2}\Big[l(U^1 - Y^1)^2 + (U^1 - U^2)^2\Big] \\ &= \mathbb{E}_{\gamma^1,\gamma^2}\Big[l(u^{1,*} - Y^1)^2\Big] + \mathbb{E}_{\gamma^1,\gamma^2}\Big[(U^1 - U^2)^2 \Big| U^1 = u^{1,*}\Big] < \infty. \end{aligned}$$

Therefore, Assumption 8.2-(e) holds.

Note that for the $\mathbb{P}_{l,n} \otimes \mathbb{P}_{l,n}$-uniform integrability condition, it is sufficient to consider compact sets of the form $[-M, M]^2$ for $M \in \mathbb{R}_+$. Let $M \in \mathbb{R}_+$. We have

$$
\begin{aligned}
w_1(y^1) &:= \sup_{(u^1,u^2)\in[-M,M]^2} l(u^1 - y^1)^2 + (u^2 - u^1)^2 \\
&= l\big(M + |y^1|\big)^2 + 4M^2
\end{aligned}
$$

and

$$
\begin{aligned}
w_2(y^2) := \sup_{(u^1,u^2)\in[-M,M]^2} f(u^1, y^2) &\leq \sup_{(u^1,u^2)\in[-M,M]^2} \exp\{y^2 u^1\} \\
= \exp\{M|y^2|\}.
\end{aligned}
$$

For functions w_1 and w_2, we have

$$
\begin{aligned}
&\lim_{R\to\infty} \sup_{n,l} \int_{\{w_1(y^1)>R\}} w_1(y^1)\, d\mathbb{P}_{l,n} \\
&\quad = \lim_{R\to\infty} \sup_{n,l} \int_{\{w_1(y^1)>R\}} w_1(q_{l,n}(y^1))\, d\mathbb{P}_{\mathrm{g}} \\
&\quad \leq \lim_{R\to\infty} \int_{\{w_1(y^1)>R\}} \Big[l\big(M + (|y^1| + 1)^2\big) + 4M^2 \Big]\, d\mathbb{P}_{\mathrm{g}} = 0
\end{aligned} \tag{9.3}
$$

and

$$
\begin{aligned}
&\lim_{R\to\infty} \sup_{n,l} \int_{\{w_2(y^2)>R\}} w_2(y^2)\, d\mathbb{P}_{l,n} \\
&\quad = \lim_{R\to\infty} \sup_{n,l} \int_{\{w_2(y^2)>R\}} w_2(q_{l,n}(y^2))\, d\mathbb{P}_{\mathrm{g}} \\
&\quad \leq \lim_{R\to\infty} \int_{\{w_2(y^2)>R\}} \exp M(|y^2| + 1)\, d\mathbb{P}_{\mathrm{g}} = 0,
\end{aligned} \tag{9.4}
$$

where (9.3) and (9.4) follow from the fact that $q_{l,n}(\mathbb{R} \setminus L) = 0$ and the integrability of w_1 and w_2 with respect to the $\mathbb{P}_{\mathrm{g}}$. By Lemma 9.1, the product $w_1 w_2$ is $\mathbb{P}_{l,n} \otimes \mathbb{P}_{l,n}$-uniformly integrable. Therefore, $\sup_{(u^1,u^2)\in[-M,M]^2} \tilde{c}(y^1, y^2, u^1, u^2)$ is also $\mathbb{P}_{l,n} \otimes \mathbb{P}_{l,n}$-uniformly integrable. Since M is arbitrary, this completes the proof. □

Proposition 9.3 and Theorem 8.14 imply that Theorems 8.11 and 8.13 are applicable to Witsenhausen's counterexample. The theorem below is the main result of this section. It states that to compute a near optimal strategy for Witsenhausen's counterexample, it is sufficient to compute an optimal strategy for the problem with finite observations obtained through uniform quantization of the observation spaces.

Theorem 9.4 *For any $\varepsilon > 0$, there exists $(l, n(l))$ and $m \in \mathbb{R}_+$ such that an optimal policy (π^{1*}, π^{2*}) in the set $\Pi_{l,n(l)}^{1,M} \times \Pi_{l,n(l)}^{2,M}$ for the cost $J_{l,n(l)}$ is ε-optimal for Witsenhausen's counterexample when (π^{1*}, π^{2*}) is extended to $\mathsf{Y}^1 \times \mathsf{Y}^2$ via $\gamma^i = \pi^{i*} \circ q_{l,n(l)}$, $i = 1, 2$, where $M := [-m, m]$ and $\Pi_{l,n(l)}^{i,M} := \{\pi^i \in \Pi_{l,n(l)}^i : \pi^i(\mathsf{Z}_{l,n}) \subset M\}$.*

9.2.3 Asymptotic Optimality of Finite Models for Witsenhausen's Counterexample

Consider the finite *observation* models introduced in Section 9.2.2 that approximate Witsenhausen's counterexample. For any $m \in \mathbb{R}_+$ and $k \in \mathbb{R}_+$, let $\mathsf{q}_{m,k} : M \to \{u_1, \ldots, u_k\}$ denote the uniform quantizer with k output levels (recall that $M := [-m, m]$). Note that $\mathsf{q}_{m,k}$ is a quantizer applied to subsets of action spaces $\mathsf{U}^i = \mathbb{R}$, $i = 1, 2$ (not to be confused with $q_{l,n}$ in Section 9.2.2). Theorem 8.15 implies that for each (l, n) and $\varepsilon > 0$, there exists an $m \in \mathbb{R}_+$ and $k \in \mathbb{R}_+$ such that

$$\inf_{\underline{\pi} \in \boldsymbol{\Pi}_{l,n}^{m,k}} J_{l,n}(\underline{\pi}) < J_{l,n}^* + \varepsilon,$$

where $\boldsymbol{\Pi}_{l,n}^{m,k} := \big\{\underline{\pi} \in \boldsymbol{\Pi}_{l,n} : \pi^i(\mathsf{Z}_{l,n}^i) \subset \mathsf{U}_{m,k}, i = 1, 2\big\}$ and $\mathsf{U}_{m,k} = \{u_1, \ldots, u_k\}$ is the set of output levels of $\mathsf{q}_{m,k}$.

Therefore, to compute a near optimal strategy for Witsenhausen's counterexample, it is sufficient to compute an optimal strategy for the finite model that is obtained through uniform quantization of observation and action spaces (i.e., $\mathbb{R}$) on finite grids when the number of grid points is sufficiently large. In particular, through constructing the uniform quantization so that both the *granular region* and the *granularity* of the quantizers are successively refined (that is the partitions generated by the quantizers are successively nested), we have the following proposition which lends itself to a numerical algorithm.

Theorem 9.5 *There exists a sequence of finite models obtained through a successive refinement of the measurement and action set partitions generated by uniform quantizers whose optimal costs will converge to the cost of Witsenhausen's counterexample.*

Remark 9.6 One may need very large computational power to find optimal solutions of the finite models constructed for Witsenhausen's counterexample since this problem was shown to be NP-complete [99]. Nonetheless, the specifics of Witsenhausen's counterexample may significantly reduce the computational complexity. Indeed, for numerical implementation one could further enhance the construction.

Observe that Agent 2 will always run a conditional estimator for Agent 1's action; that is, an optimal policy is always given by $\gamma^2(Y^2) = E[U^1|Y^2]$. Furthermore, Witsenhausen [138] has shown that an optimal policy for Agent 1 is monotone nondecreasing. These facts suggest that while searching over quantized policies, one can only focus on Agent 1, and apply quantized policies which satisfy a monotone behavior. Numerical results suggest that through such an approach one can obtain results that are quite competitive with those that exist in the literature. Furthermore, other iterative methods can be applied as in, e.g., [79, 94]. The findings in this chapter validate such efforts and show that solutions obtained through carefully designed approximation methods are guaranteed to converge to optimal solutions.

9.3 The Gaussian Relay Channel Problem

9.3.1 The Gaussian Relay Channel Problem and Its Static Reduction

An important dynamic team problem which has attracted interest is the Gaussian relay channel problem depicted in Figure. 9.2 [88, 152]. Here, Agent 1 observes a noisy version of the state X which has Gaussian distribution with zero mean and variance σ_x^2; that is, $Y^1 := X + V^0$ where V^0 is a zero mean and variance σ_0^2 Gaussian noise independent of X. Agent 1 decides its strategy U^1 based on Y^1. For $i = 2, \ldots, N$, Agent i receives $Y^i := U^{i-1} + V^{i-1}$ (a noisy version of the decision U^{i-1} of Agent $i-1$), where V^{i-1} is zero mean and variance σ_{i-1}^2 Gaussian noise independent of $\{X, V^1, \ldots, V^{i-2}, V^i, \ldots, V^{N-1}\}$, and decides its strategy U^i.

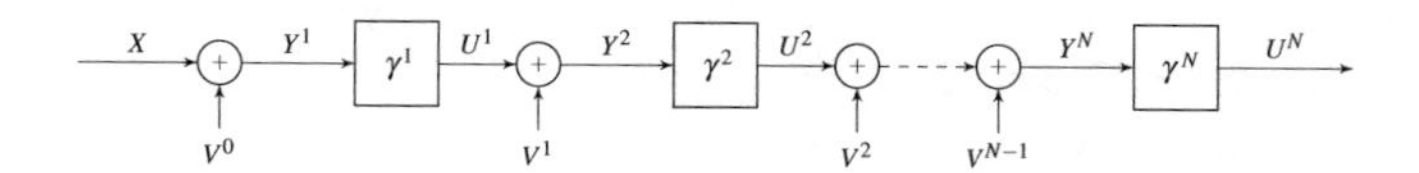

Fig. 9.2 Gaussian relay channel

The cost function of the team is given by

$$c(x, \boldsymbol{u}) := \left(u^N - x\right)^2 + \sum_{i=1}^{N-1} l_i \left(u^i\right)^2,$$

where $l_i \in \mathbb{R}_+$ for all $i = 1, \ldots, N-1$. To ease the notation, we simply take $\sigma_x = \sigma_0 = \sigma_1 = \ldots = \sigma_{N-1} = 1$. Recall that $g(y) := \frac{1}{\sqrt{2\pi}} \exp\{-y^2/2\}$. Then we have

$$W^1(S|x) = \int_S g(y^1 - x) m(dy^1)$$

$$W^i(S|u^{i-1}) = \int_S g(y^i - u^{i-1}) m(dy^i), \text{ for } i = 2, \ldots, N.$$

Recall also that $g(y-u) = f(u,y)\frac{1}{\sqrt{2\pi}}\exp\{-(y)^2/2\}$, where $f(u,y)$ is defined in (9.1). Then, for any policy $\underline{\gamma}$, we have

$$\begin{aligned} J(\underline{\gamma}) &= \int_{\mathsf{X}\times\mathsf{Y}} c(x,\boldsymbol{u}) \mathbb{P}(dx, d\boldsymbol{y}) \\ &= \int_{\mathsf{X}\times\mathsf{Y}} c(x,\boldsymbol{u}) \biggl[f(x,y^1) \prod_{i=2}^{N} f(u^{i-1}, y^i) \biggr] \mathbb{P}_{\mathsf{g}}^{N+1}(dx, d\boldsymbol{y}), \end{aligned}$$

where $\mathbb{P}_{\mathsf{g}}^{N+1}$ denotes the product of $N+1$ zero mean and unit variance Gaussian distributions. Therefore, in the static reduction of Gaussian relay channel, we have the components $\tilde{c}(x,\boldsymbol{y},\boldsymbol{u}) := c(x,\boldsymbol{u})\bigl[f(x,y^1)\prod_{i=2}^{N} f(u^{i-1},y^i)\bigr]$ and $\widetilde{\mathbb{P}}(dx,d\boldsymbol{y}) = \mathbb{P}_{\mathsf{g}}^{N+1}(dx,d\boldsymbol{y})$. Analogous to Witsenhausen's counterexample, the agents observe independent zero mean and unit variance Gaussian random variables.

9.3.2 Approximation of the Gaussian Relay Channel Problem

In this subsection, the approximation problem for the Gaussian relay channel is considered using the static reduction formulation. Analogous to Section 9.2, we prove that the conditions of Theorem 8.14 hold for Gaussian relay channel, and so Theorems 8.11 and 8.13 can be applied.

The cost function of the static reduction is given by

$$\tilde{c}(x,\boldsymbol{y},\boldsymbol{u}) := c(x,\boldsymbol{u})\bigl[f(x,y^1)\prod_{i=2}^{N} f(u^{i-1},y^i)\bigr],$$

where $f(u,y)$ is given in (9.1).

Recall the uniform quantizer $q_{l,n}$ on $L := [l,-l]$ having n output levels from Section 9.2. We extend $q_{l,n}$ to $\mathbb{R}$ by mapping $\mathbb{R}\setminus L$ to $y_0 = 0$. Recall also the set $\mathsf{Z}_{l,n} := \{y_0, y_1, \ldots, y_n\}$ and the probability measure $\mathbb{P}_{l,n}$ on $\mathsf{Z}_{l,n}$ given by

$$\mathbb{P}_{l,n}(y_i) = \mathbb{P}_{\mathsf{g}}(q_{l,n}^{-1}(y_i)).$$

Define $\Pi_{l,n}^i := \{\pi^i : \mathsf{Z}_{l,n} \to \mathsf{U}^i,\ \pi^i \text{ measurable}\}$ and

$$J_{l,n}(\underline{\pi}) := \int_{\mathsf{X}} \sum_{\mathbf{y} \in \prod_{i=1}^{N} \mathsf{Z}_{l,n}} \tilde{c}(x, \mathbf{y}, \underline{\pi}(\mathbf{y})) \prod_{i=1}^{N} \mathbb{P}_{l,n}(y_i) \mathbb{P}_{\mathsf{g}}(dx),$$

where $\underline{\pi} := (\pi^1, \ldots, \pi^N)$. Define $\tilde{\mathbb{P}}_{l,n}(dx, d\mathbf{y}) := \prod_{i=1}^{N} \mathbb{P}_{l,n}(dy_i) \mathbb{P}_{\mathsf{g}}(dx)$.

Proposition 9.7 *The Gaussian relay channel problem satisfies the conditions in Theorem 8.14.*

Proof It is clear that Assumption 8.2-(a),(b),(c) and Assumption 7.1 hold. For any strategy $\underline{\gamma}$, let $\mathbb{E}_{\underline{\gamma}}$ denote the corresponding expectation operation. To prove Assumption 8.2-(e), pick $\underline{\gamma}$ with $J(\underline{\gamma}) < \infty$. Analogous to Lemma 9.2, one can prove that $\mathbb{E}_{\underline{\gamma}}\big[(U^N)^2\big] < \infty$. For any $i = 1, \ldots, N$, by the law of total expectation we can write

$$J(\underline{\gamma}) = \mathbb{E}_{\underline{\gamma}}\Big[\mathbb{E}_{\underline{\gamma}}\big[c(X, \boldsymbol{U}) \big| U^i\big]\Big]$$

and

$$\mathbb{E}_{\underline{\gamma}}\Big[(U^N)^2\Big] = \mathbb{E}_{\underline{\gamma}}\Big[\mathbb{E}_{\underline{\gamma}}\big[(U^N)^2 \big| U^i\big]\Big].$$

Therefore, for each $i = 1, \ldots, N$, there exists $u^{i,*} \in \mathsf{U}^i$ such that $\mathbb{E}_{\underline{\gamma}}\big[c(X, \boldsymbol{U}) \big| U^i = u^{i,*}\big] < \infty$ and $\mathbb{E}_{\underline{\gamma}}\big[(U^N)^2 \big| U^i = u^{i,*}\big] < \infty$. Then we have

$$\begin{aligned}
J(\underline{\gamma}^{-i}, \gamma^i_{u^{i,*}}) &= \mathbb{E}_{\underline{\gamma}^{-i}, \gamma^i_{u^{i,*}}}\Big[(U^N - X)^2 + \sum_{j=1}^{N-1} l_j (U^j)^2\Big] \\
&\leq \mathbb{E}_{\underline{\gamma}^{-i}, \gamma^i_{u^{i,*}}}\Big[2X^2 + 2(U^N)^2 + \sum_{j=1}^{N-1} l_j (U^j)^2\Big] \\
&= \mathbb{E}_{\underline{\gamma}}\Big[\sum_{j=1}^{i-1} l_j (U^j)^2 + 2X^2\Big] \\
&\quad + \mathbb{E}_{\underline{\gamma}}\Big[2(U^N)^2 + \sum_{j=i}^{N-1} l_j (U^j)^2 \Big| U^i = u^{i,*}\Big] < \infty.
\end{aligned}$$

Therefore, Assumption 8.2-(e) holds.

Analogous to the proof of Proposition 9.3, for the $\tilde{\mathbb{P}}_{l,n}$-uniform integrability condition it is sufficient to consider compact sets of the form $[-M, M]^N$ for $M \in \mathbb{R}_+$. We have

$$w_1(x) := \sup_{\boldsymbol{u} \in [-M,M]^N} \left(u^N - x\right)^2 + \sum_{i=1}^{N-1} l_i \left(u^i\right)^2$$
$$= \left(M + x\right)^2 + \sum_{i=1}^{N-1} l_i M^2$$

and

$$\prod_{i=2}^{N} w_{i,2}(y^i) := \sup_{\boldsymbol{u} \in [-M,M]^N} \prod_{i=2}^{N} f(u^{i-1}, y^i) \quad \leq \quad \sup_{\boldsymbol{u} \in [-M,M]^N} \prod_{i=2}^{N} \exp\{y^i u^{i-1}\}$$
$$= \prod_{i=2}^{N} \exp\{M|y^i|\}.$$

It can be shown as in the proof of Proposition 9.3 that each $w_{i,2}$ is $\mathbb{P}_{l,n}$-uniformly integrable. Letting $\tilde{w}_1(x, y^1) := w_1(x) f(x, y^1)$, we have

$$\int_{\mathsf{X}\times\mathsf{Y}^1} \tilde{w}_1(x, y^1) \exp\{|x|\}\, d\mathbb{P}_{\mathsf{g}}^2 = \int_{\mathsf{X}} w_1(x) \exp\{|x|\}\, d\mathbb{P}_{\mathsf{g}} < \infty$$

and therefore

$$\lim_{R\to\infty} \sup_{l,n} \int_{\{\tilde{w}_1 > R\}} \tilde{w}_1(x, y^1)\, d\mathbb{P}_{l,n} d\mathbb{P}_{\mathsf{g}}$$
$$= \lim_{R\to\infty} \sup_{l,n} \int_{\{\tilde{w}_1 > R\}} \tilde{w}_1(x, q_{l,n}(y^1))\, d\mathbb{P}_{\mathsf{g}}^2$$
$$\leq \lim_{R\to\infty} \int_{\{\tilde{w}_1 > R\}} w_1(x) \exp\left\{\frac{-x^2 + 2xy^1 + 2|x|}{2}\right\} d\mathbb{P}_{\mathsf{g}}^2$$
$$= \lim_{R\to\infty} \int_{\{\tilde{w}_1 > R\}} \tilde{w}_1(x, y^1) \exp\{|x|\}\, d\mathbb{P}_{\mathsf{g}}^2 = 0.$$

Hence, by Lemma 9.1, the product $\tilde{w}_1 \prod_{i=2}^{N} w_{i,2}$ is $\tilde{\mathbb{P}}_{l,n}$-uniformly integrable. Therefore, $\sup_{\boldsymbol{u} \in [-M,M]^N} \tilde{c}(x, \boldsymbol{y}, \boldsymbol{u})$ is also $\tilde{\mathbb{P}}_{l,n}$-uniformly integrable. Since M is arbitrary, this completes the proof. □

The preceding proposition and Theorem 8.14 imply, via Theorems 8.11 and 8.13, that an optimal strategy for Gaussian relay channel can be approximated by strategies obtained from finite models. The following theorem is the main result of this section.

Theorem 9.8 *For any* $\varepsilon > 0$, *there exists* $(l, n(l))$ *and* $m \in \mathbb{R}_+$ *such that an optimal policy* $\underline{\pi}^*$ *in the set* $\prod_{i=1}^{N} \Pi_{l,n(l)}^{i,M}$ *for the cost* $J_{l,n(l)}$ *is* ε*-optimal for Gaussian relay channel when* $\underline{\pi}^*$ *is extended to* Y *via* $\gamma^i = \pi^{i*} \circ q_{l,n(l)}$, $i = 1, \ldots, N$, *where* $M := [-m, m]$ *and* $\Pi_{l,n(l)}^{i,M} := \{\pi^i \in \Pi_{l,n(l)}^i : \pi^i(\mathsf{Z}_{l,n}) \subset M\}$.

9.3.3 Asymptotic Optimality of Finite Models for Gaussian Relay Channel Problem

Recall that for any $m \in \mathbb{R}_+$ and $k \in \mathbb{R}_+$, $\mathsf{q}_{m,k} : M \to \{u_1, \ldots, u_k\}$ denote the uniform quantizer with k output levels where $M := [-m, m]$. The quantizer $\mathsf{q}_{m,k}$ is applied to subsets of action spaces $\mathsf{U}^i = \mathbb{R}$, $i = 1, 2$, and so it should not be confused with $q_{l,n}$ in Section 9.3.2. By Theorem 8.15, for each (l, n) and $\varepsilon > 0$ there exists an $m \in \mathbb{R}_+$ and $k \in \mathbb{R}_+$ such that

$$\inf_{\underline{\pi} \in \boldsymbol{\Pi}_{l,n}^{m,k}} J_{l,n}(\underline{\pi}) < J_{l,n}^* + \varepsilon,$$

where $\boldsymbol{\Pi}_{l,n}^{m,k} := \left\{\underline{\pi} \in \boldsymbol{\Pi}_{l,n} : \pi^i(\mathsf{Z}_{l,n}^i) \subset \mathsf{U}_{m,k}, i = 1, \ldots, N\right\}$ and $\mathsf{U}_{m,k} = \{u_1, \ldots, u_k\}$ is the set of output levels of $\mathsf{q}_{m,k}$.

Hence, we can compute a near optimal strategy for Gaussian relay channel problem by computing an optimal strategy for the finite model that is obtained through uniform quantization of observation and action spaces (i.e., the real line) on finite grids when the number of grid points is sufficiently large.

Theorem 9.9 *There exists a sequence of finite models obtained through a successive refinement of the measurement and action set partitions generated by uniform quantizers whose optimal costs will converge to the cost of the Gaussian relay channel problem.*

9.4 Concluding Remarks

In this chapter, using the results from the previous chapter, approximation results were established for the well-known counterexample of Witsenhausen and Gaussian relay channel problem. Our approximation approach to Witsenhausen's counterexample thus provides, to our knowledge, the first rigorously established result that for any $\varepsilon > 0$, one can construct an ε-optimal strategy through an explicit solution of a conceptually simpler problem. The results of this chapter build on [120].

For a further reading on approximation of Witsenhausen's counterexample, we refer the reader to [9, 10, 53, 86, 87, 93, 94]. With the exception of [53], these works are mostly experimental; that is, they do not rigorously prove the convergence of approximate solutions. In [53], both the error bound and the near optimal solutions depend on the knowledge of the optimal strategy for Witsenhausen's counterexample. Indeed, the method devised in [53] implicitly corresponds to the discretization of only the action spaces of the agents.

References

1. D. Adelman, D. Klabjan, Computing near-optimal policies in generalized joint replenishment. INFORMS J. Comput. **24**(1), 148–164 (2012)
2. C.D. Aliprantis, K.C. Border, *Infinite Dimensional Analysis* (Springer, Berlin, 2006)
3. E. Altman, *Constrained Markov Decision Processes* (Chapman and Hall/CRC, Boca Raton, FL, 1999)
4. M. Andersland, D. Teneketzis, Information structures, causality, and non-sequential stochastic control, I: design-independent properties. SIAM J. Control Optim. **30**, 1447–1475 (1992)
5. M. Andersland, D. Teneketzis, Information structures, causality, and non-sequential stochastic control, II: design-dependent properties. SIAM J. Control Optim. **32**, 1726–1751 (1994)
6. A. Arapostathis, V.S. Borkar, E. Fernandez-Gaucherand, M.K. Ghosh, S.I. Marcus, Discrete-time controlled Markov processes with average cost criterion: a survey. SIAM J. Control Optim. **31**, 282–344 (1993)
7. A. Arapostathis, V.S. Borkar, M.K. Ghosh, *Ergodic Control of Diffusion Processes*, vol. 143 (Cambridge University Press, 2012)
8. K.J. Arrow, The informational structure of the firm. Am. Econ. Rev. **75**, 303–307 (1985)
9. M. Baglietto, T. Parisini, R. Zoppoli, Numerical solutions to the Witsenhausen counterexample by approximating networks. IEEE Trans. Autom. Control **46**(9), 1471–1477 (2001)
10. R. Bansal, T. Başar, Stochastic team problems with nonclassical information revisited: When is an affine law optimal? IEEE Trans. Autom. Control **32**, 554–559 (1987)
11. A. Barnivok, *A Course in Convexity* (American Mathematical Society, Providence, 2002)
12. C.L. Beck, S. Lall, T. Liang, M. West, Model reduction, optimal prediction, and the Mori-Zwanzig representation of Markov chains, in *CDC 2009*, Shanghai (2009), pp. 3282–3287
13. D.P. Bertsekas, Convergence of discretization procedures in dynamic programming. IEEE Trans. Autom. Control **20**(3), 415–419 (1975)
14. D.P. Bertsekas, *Dynamic Programming and Optimal Control: Volume II* (Athena Scientific, Belmont, 1995)
15. D. Bertsekas, *Nonlinear Programming* (Athena Scientific, Belmont, MA, 1999)
16. D.P. Bertsekas, S.E. Shreve, *Stochastic Optimal Control: The Discrete Time Case* (Academic, New York, 1978)
17. D.P. Bertsekas, J.N. Tsitsiklis, *Neuro-Dynammic Programming* (Athena Scientific, Belmont, 1996)
18. A.G. Bhatt, A. Budhiraja, R.L. Karandikar, Markov property and ergodicity of the nonlinear filter. SIAM J. Control Optim. **39**(3), 928–949 (2000)
19. P. Billingsley, *Probability and Measure*, 3rd edn. (Wiley, New York, 1995)

N. Saldi et al., *Finite Approximations in Discrete-Time Stochastic Control*,
Systems & Control: Foundations & Applications,
https://doi.org/10.1007/978-3-319-79033-6

20. P. Billingsley, *Convergence of Probability Measures*, 2nd edn. (Wiley, New York, 1999)
21. D. Blackwell, D. Freedman, M. Orkin, The optimal reward operator in dynamic programming. Ann. Probab. **2**(2), 926–941 (1974)
22. G. Böcherer, B.C Geiger, Optimal quantization for distribution synthesis. IEEE Trans. Inf. Theory **62**(11), 6162–6172 (2016)
23. V.I. Bogachev, *Measure Theory: Volume II* (Springer, Berlin, 2007)
24. V. Borkar, Convex analytic methods in Markov decision processes, in *Handbook of Markov Decision Processes* ed. by E.A. Feinberg, A. Shwartz (Kluwer Academic, Boston, 2002)
25. A. Brooks, S. Williams, A Monte Carlo update for parametric POMDPs, in *Proceedings of International Symposium on Robotics Research* (2007)
26. A. Brooks, A. Makarenko, S. Williams, H. Durrant-Whyte, Parametric POMDPs for planning in continuous state spaces. Robot. Auton. Syst. **54**(11), 887–897 (2006)
27. A. Budhiraja, On invariant measures of discrete time filters in the correlated signal-noise case. Ann. Appl. Probab. **12**(3), 1096–1113 (2002)
28. L. Busoniu, D. Ernst, B.D. Schutter, R. Babuska, Approximate dynamic programming with a fuzzy parametrization. Automatica **46**, 804–814 (2010)
29. P.E. Caines, *Linear Stochastic Systems* (Wiley, New York, NY, 1988)
30. R. Cavazos-Cadena, Finite-state approximations for denumerable state discounted Markov decision processes. Appl. Math. Optim. **14**, 1–26 (1986)
31. C.D. Charalambous, N.U. Ahmed, Equivalence of decentralized stochastic dynamic decision systems via Girsanov's measure transformation, in *IEEE Conference on Decision and Control (CDC)* (IEEE, Los Angeles, 2014), pp. 439–444
32. S.B. Connor, G. Fort, State-dependent Foster-Lyapunov criteria for subgeometric convergence of Markov chains. Stoch. Process Appl. **119**, 176–4193 (2009)
33. T.M. Cover, J.A. Thomas, *Elements of Information Theory*, 2nd edn. (Wiley, New York, 2006)
34. D.P. de Farias, B.V. Roy, The linear programming approach to approximate dynamic programming. Oper. Res. **51**(6), 850–865 (2003)
35. D.P. de Farias, B.V. Roy, On constraint sampling in the linear programming approach to approximate dynamic programming. Math. Oper. Res. **29**(3), 462–478 (2004)
36. B. de Saporta, F. Dufour, C. Nivot, Partially observed optimal stopping problem for discrete-time Markov processes. 4OR **15**(3), 277–302 (2017)
37. P.R. De Waal, J.H. Van Schuppen, A class of team problems with discrete action spaces: optimality conditions based on multimodularity. SIAM J. Control Optim. **38**(3), 875–892 (2000)
38. R. Douc, G. Fort, E. Moulines, P. Soulier, Practical drift conditions for subgeometric rates of convergence. Ann. Appl. Probab **14**, 1353–1377 (2004)
39. R.M. Dudley, *Real Analysis and Probability*, 2nd edn. (Cambridge University Press, Cambridge, 2002)
40. F. Dufour, T. Prieto-Rumeau, Approximation of Markov decision processes with general state space. J. Math. Anal. Appl. **388**, 1254–1267 (2012)
41. F. Dufour, T. Prieto-Rumeau, Finite linear programming approximations of constrained discounted Markov decision processes. SIAM J. Control Optim. **51**(2), 1298–1324 (2013)
42. F. Dufour, T. Prieto-Rumeau, Approximation of average cost Markov decision processes using empirical distributions and concentration inequalities. Stochastics **87**, 273–307 (2014)
43. F. Dufour, T. Prieto-Rumeau, Stochastic approximations of constrained discounted Markov decision processes. J. Math. Anal. Appl. **413**, 856–879 (2014)
44. E.B. Dykin, *Controlled Markov Processes* (Springer, Berlin/New York, 1979)
45. P.M. Esfahani, T. Sutter, D. Kuhn, J. Lygeros, From infinite to finite programs: explicit error bounds with applications to approximate dynamic programming. arXiv:1701.06379 (2017)
46. A.M. Farahmand, R. Munos, C. Szepesvari, Error propagation for approximate policy and value iteration, in *Advances in Neural Information Processing Systems* (2010), pp. 568–576

47. E.A. Feinberg, On measurability and representation of strategic measures in Markov decision processes. Stat. Probab. Game Theory **30**, 29–43 (1996)
48. E.A. Feinberg, M.E. Lewis, Optimality inequalities for average cost Markov decision processes and the stochastic cash balance problem. Math. Oper. Res. **32**(4), 769–783 (2007)
49. E.A. Feinberg, P.O. Kasyanov, N.V. Zadioanchuk, Average cost Markov decision processes with weakly continuous transition probabilities. Math. Oper. Res. **37**(4), 591–607 (2012)
50. E.A. Feinberg, P.O. Kasyanov, M.Z. Zgurovsky, Partially observable total-cost Markov decision process with weakly continuous transition probabilities. Math. Oper. Res. **41**, 656–681 (2016)
51. B.L. Fox, Finite-state approximations to denumerable state dynamic programs. J. Math. Anal. Appl. **34**, 665–670 (1971)
52. G. Gnecco, M. Sanguinetti, Suboptimal solutions to network team optimization problems, in *INOC 2009*, Pisa (2009)
53. G. Gnecco, M. Sanguinetti, New insights into Witsenhausen's counterexample. Optim. Lett. **6**, 1425–1446 (2012)
54. G. Gnecco, M. Sanguinetti, M. Gaggero, Suboptimal solutions to team optimization problems with stochastic information structure. SIAM J. Control Optim. **22**(1), 212–243 (2012)
55. E. Gordienko, O. Hernandez-Lerma, Average cost Markov control processes with weighted norms: existence of canonical policies. Appl. Math. **23**(2), 199–218 (1995)
56. A. Granas, J. Dugundji, *Fixed Point Theory* (Springer, New York, 2003)
57. R.M. Gray, *Entropy and Information Theory* (Springer, New York, 1990)
58. R.M. Gray, D.L. Neuhoff, Quantization. IEEE Trans. Inf. Theory **44**, 2325–2383 (1998)
59. P. Grover, S.Y. Park, A. Sahai, Approximately optimal solutions to the finite-dimensional Witsenhausen counterexample. IEEE Trans. Autom. Control **58**(9), 2189–2204 (2013)
60. A. Gupta, S. Yüksel, T. Basar, C. Langbort, On the existence of optimal policies for a class of static and sequential dynamic teams. SIAM J. Control Optim. **53**(3), 1681–1712 (2015)
61. L. Györfi, M. Kohler, Nonparametric estimation of conditional distributions. IEEE Trans. Inf. Theory **53**, 1872–1878 (2007)
62. A. Gyorgy, T. Linder, Codecell convexity in optimal entropy-constrained vector quantization. IEEE Trans. Inf. Theory **49**(7), 1821–1828 (2003)
63. M. Hairer, *Ergodic Properties of Markov Processes*. Lecture Notes (2006)
64. M. Hairer, *Convergence of Markov Processes*. Lecture Notes (University of Warwick, 2010)
65. O. Hernández-Lerma, Finite-state approximations for denumerable multidimensional state discounted Markov decision processes. J. Math. Anal. Appl. **113**, 382–388 (1986)
66. O. Hernández-Lerma, *Adaptive Markov Control Processes* (Springer, New York, 1989)
67. O. Hernandez-Lerma, J. Gonzalez-Hernandez, Constrained Markov control processes in Borel spaces:the discounted case. Math. Methods Oper. Res. **52**, 271–285 (2000)
68. O. Hernández-Lerma, J.B. Lasserre, *Discrete-Time Markov Control Processes: Basic Optimality Criteria* (Springer, Berlin, 1996)
69. O. Hernández-Lerma, J.B. Lasserre, *Further Topics on Discrete-Time Markov Control Processes* (Springer, New York, 1999)
70. O. Hernández-Lerma, J.B. Lasserre, *Markov Chains and Invariant Probabilities* (Birkhauser, Basel, 2003)
71. O. Hernández-Lerma, R. Romera, Limiting discounted-cost control of partially observable stochastic systems. SIAM J. Control Optim. **40**(2), 348–369 (2001)
72. O. Hernández-Lerma, R. Montes-De-Oca, R. Cavazos-Cadena, Recurrence conditions for Markov decision processes with Borel state space: a survey. Ann. Oper. Res. **28**(1), 29–46 (1991)
73. O. Hernandez-Lerma, J. Gonzalez-Hernandez, R.R. Lopez-Martinez, Constrained average cost Markov control processes in Borel spaces. SIAM J. Control Optim. **42**(2), 442–468 (2003)
74. P. Hernández-Lerma, J.B. Lasserre, Error bounds for rolling horizon policies in discrete-time Markov control processes. IEEE Trans. Autom. Control **35**(10), 1118–1124 (1990)

75. K. Hinderer, Lipshitz continuity of value functions in Markovian desision processes. Math. Methods Oper. Res. **62**, 3–22 (2005)
76. Y.C. Ho, K.C. Chu, Team decision theory and information structures in optimal control problems - part I. IEEE Trans. Autom. Control **17**, 15–22 (1972)
77. Y.C. Ho, K.C. Chu, On the equivalence of information structures in static and dynamic teams. IEEE Trans. Autom. Control **18**(2), 187–188 (1973)
78. A. Jaśkiewicz, A.S. Nowak, On the optimality equation for average cost Markov control processes with Feller transition probabilities. J. Math. Anal. Appl. **316**, 495–509 (2006)
79. J. Karlsson, A. Gattami, T.J. Oechtering, M. Skoglund, Iterative source-channel coding approach to Witsenhausen's counterexample, in *ACC 2011*, San Francisco (2011)
80. J.C. Krainak, J.L. Speyer, S.I. Marcus, Static team problems – part I: sufficient conditions and the exponential cost criterion. IEEE Trans. Autom. Control **27**, 839–848 (1982)
81. W. Kreitmeier, Optimal vector quantization in terms of Wasserstein distance. J. Multivar. Anal. **102**(8), 1225–1239 (2011)
82. V. Krishnamurthy, *Partially Observed Markov Decision Processes: From Filtering to Controlled Sensing* (Cambridge University Press, Cambridge, 2016)
83. K. Kuratowski, *Topology: Volume I* (Academic, New York, 1966)
84. H.J. Kushner, *Introduction to Stochastic Control Theory* (Holt, Rinehart and Winston, New York, 1972)
85. H. Langen, Convergence of dynamic programming models. Math. Oper. Res. **6**(4), 493–512 (1981)
86. J.T. Lee, E. Lau, Y. Ho, The Witsenhausen counterexample: a hierarchical search approach for nonconvex optimization problems. IEEE Trans. Autom. Control **46**(3), 382–397 (2001)
87. N. Li, R. Marden, J.S. Shamma, Learning approach to the Witsenhausen counterexample from a view of potential games, in *CDC 2009*, Shanghai (2009), pp. 157–162
88. G.M. Lipsa, N.C. Martins, Optimal memoryless control in Gaussian noise: a simple counterexample. Automatica **47**, 552–558 (2011)
89. W.S. Lovejoy, Computationally feasible bounds for partially observed Markov decision processes. Oper. Res. **39**(1), 162–175 (1991)
90. W.S. Lovejoy, A survey of algorithmic methods for partially observed Markov decision processes. Ann. Oper. Res. **28**, 47–66 (1991)
91. A.S. Manne, Linear programming and sequential decisions. Manag. Sci. **6**(3), 259–267 (1980)
92. J. Marschak, Elements for a theory of teams. Manag. Sci. **1**, 127–137 (1955)
93. W.M. McEneaney, S.H. Han, Optimization formulation and monotonic solution method for the Witsenhausen problem. Automatica **55**, 55–65 (2015)
94. M. Mehmetoglu, E. Akyol, K. Rose, A deterministic annealing approach to Witsenhausen's counterexample, in *ISIT 2014*, Honolulu (2014)
95. S.P. Meyn, *Control Techniques for Complex Networks* (Cambridge University Press, Cambridge, 2007)
96. S.P. Meyn, R.L. Tweedie, *Markov Chains and Stochastic Stability* (Springer, New York, 1993)
97. S. Mitter, A. Sahai, Information and control: Witsenhausen revisited, in *Learning, Control, and Hybrid Systems* (1999), pp. 281–293
98. R. Ortner, Pseudometrics for state aggregation in average reward Markov decision processes, in *Algorithmic Learning Theory* (Springer, Berlin, 2007)
99. C.H. Papadimitriou, J. Tsitsiklis, Intractable problems in control theory. SIAM J. Control Optim. **24**(4), 639–654 (1986)
100. K.R. Parthasarathy, *Probability Measures on Metric Spaces* (AMS, Providence, 1967)
101. J. Pineau, G. Gordon, S. Thrun, Anytime point-based approximations for large POMDPs. J. Artif. Intell. Res. **27**, 335–380 (2006)
102. J.M. Porta, N. Vlassis, M.T.J. Spaan, P. Poupart, Point-based value iteration for continuous POMDPs. J. Mach. Learn. Res. **7**, 2329–2367 (2006)
103. W.B. Powell, *Approximate Dynamic Programming* (Wiley, Hoboken, NJ, 2007)

104. M.L. Puterman, *Markov Decision Processes* (Wiley, Hoboken, NJ, 2005)
105. R. Radner, Team decision problems. Ann. Math. Stat. **33**, 857–881 (1962)
106. Z. Ren, B.H. Krogh, State aggregation in Markov decision processes, in *CDC 2002*, Las Vegas (2002)
107. Y.A. Reznik, An algorithm for quantization of discrete probability distributions, in *DCC 2011* (2011), pp. 333–342
108. D. Rhenius, Incomplete information in Markovian decision models. Ann. Stat. **2**, 1327–1334 (1974)
109. G.O. Roberts, J.S. Rosenthal, General state space Markov chains and MCMC algorithms. Probab. Surv. **1**, 20–71 (2004)
110. B.V. Roy, Performance loss bounds for approximate value iteration with state aggregation. Math. Oper. Res. **31**(2), 234–244 (2006)
111. W.J. Runggaldier, L. Stettner, *Approximations of Discrete Time Partially Observed Control Problems*. Applied Mathematics Monographs CNR (Giardini Editori, Pisa, 1994)
112. N. Saldi, T. Linder, S. Yüksel, Approximation of stationary control policies by quantized control in Markov decision processes, in *52nd Annual Allerton Conference on Communication, Control, and Computing*, Monticello, IL (2013), pp. 78–84
113. N. Saldi, T. Linder, S. Yüksel, Asymptotic optimality of quantized policies in stochastic control under weak continuity conditions, in *IEEE Conference on Decision Control*, Los Angeles (2014)
114. N. Saldi, T. Linder, S. Yüksel, Asymtotic optimality and rates of convergence of quantized stationary policies in stochastic control. IEEE Trans. Autom. Control **60**(2), 553–558 (2015)
115. N. Saldi, T. Linder, S. Yüksel, Finite state approximations of Markov decision processes with general state and action spaces, in *American Control Conference*, Chicago (2015)
116. N. Saldi, S. Yüksel, T. Linder, Finite-state approximation of Markov decision processes with unbounded costs and Borel spaces, in *IEEE Conference Decision Control*, Japan (2015)
117. N. Saldi, S. Yüksel, T. Linder, Finite-state approximations to constrained Markov decision processes with Borel spaces, in *52nd Annual Allerton Conference on Communication, Control, and Computing*, Monticello, IL (2015), pp. 567–572
118. N. Saldi, S. Yüksel, T. Linder, Near optimality of quantized policies in stochastic control under weak continuity conditions. J. Math. Anal. Appl. **435**(1), 321–337 (2016)
119. N. Saldi, S. Yüksel, T. Linder, Asymptotic optimality of finite approximations to Markov decision processes with Borel spaces. Math. Oper. Res. **42**(4), 945–978 (2017)
120. N. Saldi, S. Yüksel, T. Linder, Finite model approximations and asymptotic optimality of quantized policies in decentralized stochastic control. IEEE Trans. Autom. Control **62**(5), 2360–2373 (2017)
121. M. Schäl, On dynamic programming: compactness of the space of policies. Stoch. Process. Appl. **3**(4), 345–364 (1975)
122. R. Serfozo, Convergence of Lebesgue integrals with varying measures. Sankhya Ser. A **44**, 380–402 (1982)
123. S.E. Shreve, D.P. Bertsekas, Universally measurable policies in dynamic programming. Math. Oper. Res. **4**(1), 15–30 (1979)
124. T. Smith, R. Simmons, Point-based POMDP algorithms: improved analysis and implementation. arXiv preprint arXiv:1207.1412 (2012)
125. D. Teneketzis, On information structures and nonsequential stochastic control. CWI Q. **9**, 241–260 (1996)
126. J.N. Tsitsiklis, B.V. Roy, Feature-based methods for large scale dynamic programming. Mach. Learn. **22**, 59–94 (1996)
127. P. Tuominen, R.L. Tweedie, Subgeometric rates of convergence of f-ergodic Markov chains. Adv. Ann. Appl. Probab. **26**(3), 775–798 (1994)
128. T. van Zandt, Decentralized information processing in the theory of organizations, in *Contemporary Economic Issues, Vol. 4: Economic Design and Behavior*, ed. by M. Sertel (MacMillan, London, 1999), pp. 125–160

129. O. Vega-Amaya, The average cost optimality equation: a fixed point approach. Bol. Soc. Mat. Mex. **9**(3), 185–195 (2003)
130. C. Villani, *Optimal Transport: Old and New* (Springer, Berlin, 2009)
131. N. Vlassis, M.T.J. Spaan, Perseus: randomized point-based value iteration for POMDPs. J. Artif. Intell. Res. **24**, 195–220 (2005)
132. C.C. White, A survey of solution techniques for the partially observed Markov decision process. Ann. Oper. Res. **32**, 215–230 (1991)
133. C.C. White, W.T. Scherer, Finite-memory suboptimal design for partially observed Markov decision processes. Oper. Res. **42**(3), 439–455 (1994)
134. D.J. White, Finite-state approximations for denumerable state infinite horizon discounted Markov decision processes. J. Math. Anal. Appl. **74**, 292–295 (1980)
135. D.J. White, Finite-state approximations for denumerable state infinite horizon discounted Markov decision processes with unbounded rewards. J. Math. Anal. Appl. **186**, 292–306 (1982)
136. W. Whitt, Approximations of dynamic programs I. Math. Oper. Res. **3**(3), 231–243 (1978)
137. W. Whitt, Approximations of dynamic programs II. Math. Oper. Res. **4**(2), 179–185 (1979)
138. H.S. Witsenhausen, A counterexample in stochastic optimum control. SIAM J. Control Optim. **6**(1), 131–147 (1968)
139. H.S. Witsenhausen, Separation of estimation and control for discrete time systems. Proc. IEEE **59**, 1557–1566 (1971)
140. H.S. Witsenhausen, On information structures, feedback and causality. SIAM J. Control **9**, 149–160 (1971)
141. H.S. Witsenhausen, A standard form for sequential stochastic control. Math. Syst. Theory **7**, 5–11 (1973)
142. H.S. Witsenhausen, The intrinsic model for discrete stochastic control: some open problems, in *Control Theory, Numerical Methods and Computer Systems Modelling* Lecture Notes in Economics and Mathematical Systems, vol. 107 (Springer, Berlin, 1975), pp. 322–335
143. H.S. Witsenhausen, Equivalent stochstic control problems. Math. Control Signal Syst. **1**(1), 3–11 (1988)
144. Y. Wu, S. Verdú, Witsenhausen's counterexample: a view from optimal transport theory, in *Proceedings of the IEEE Conference on Decision and Control, Florida, USA* (2011), pp. 5732–5737
145. Y. Yamada, S. Tazaki, R.M. Gray, Asymptotic performance of block quantizers with difference distortion measures. IEEE Trans. Inf. Theory **26**, 6–14 (1980)
146. H. Yu, D.P. Bertsekas, Discretized approximations for POMDP with average cost, in *Conference on Uncertainity in Artifical Intelligence* (2004), pp. 619–627
147. H. Yu, D.P. Bertsekas, On near optimality of the set of finite-state controllers for average cost POMDP. Math. Oper. Res. **33**(1), 1–11 (2008)
148. S Yüksel, T. Başar, *Stochastic Networked Control Systems: Stabilization and Optimization Under Information Constraints* (Springer, New York, 2013)
149. S. Yüksel, T. Linder, Optimization and convergence of observation channels in stochastic control. SIAM J. Control Optim. **50**, 864–887 (2012)
150. S. Yüksel, N. Saldi, Convex analysis in decentralized stochastic control, strategic measures and optimal solutions. SIAM J. Control Optim. **55**(1), 1–28 (2017)
151. A.A. Yushkevich, Reduction of a controlled Markov model with incomplete data to a problem with complete information in the case of Borel state and control spaces. Theory Probab. Appl. **21**, 153–158 (1976)
152. A.A. Zaidi, S. Yüksel, T. Oechtering, M. Skoglund, On optimal policies for control and estimation over a Gaussian relay channel. Automatica **49**(9), 2892–2897 (2013)

153. D. Zhang, D. Adelman, An approxiamte dynamic programming approach to network revenue management with customer choice. Transp. Sci. **43**(3), 381–394 (2009)
154. E. Zhou, M.C. Fu, S.I. Marcus, A density projection approach to dimension reduction for continuous-state POMDPs, in *47th IEEE Conference on Decision and Control, 2008. CDC 2008* (IEEE, Cancun, 2008), pp. 5576–5581
155. E. Zhou, M.C. Fu, S.I. Marcus, Solving continuous-state POMDPs via density projection. IEEE Trans. Autom. Control **55**(5), 1101–1116 (2010)
156. R. Zhou, E.A. Hansen, An improved grid-based approximation algorithm for POMDPs, in *International Joint Conference on Artificial Intelligence* (2001), pp. 707–714

Index

N. Saldi et al., *Finite Approximations in Discrete-Time Stochastic Control*,
Systems & Control: Foundations & Applications,
https://doi.org/10.1007/978-3-319-79033-6